About Island Press

ISLAND PRESS, a nonprofit organization, publishes, markets, and distributes the most advanced thinking on the conservation of our natural resources—books about soil, land, water, forests, wildlife, and hazardous and toxic wastes. These books are practical tools used by public officials, business and industry leaders, natural resource managers, and concerned citizens working to solve both local and global resource problems.

Founded in 1978, Island Press reorganized in 1984 to meet the increasing demand for substantive books on all resource-related issues. Island Press publishes and distributes under its own imprint and offers these services to other nonprofit organizations.

Support for Island Press is provided by Apple Computer, Inc., Mary Reynolds Babcock Foundation, Geraldine R. Dodge Foundation, The Energy Foundation, The Charles Engelhard Foundation, The Ford Foundation, Glen Eagles Foundation, The George Gund Foundation, William and Flora Hewlett Foundation, The Joyce Foundation, The John D. and Catherine T. MacArthur Foundation, The Andrew W. Mellon Foundation, The Joyce Mertz-Gilmore Foundation, The New-Land Foundation, The J. N. Pew, Jr., Charitable Trust, Alida Rockefeller, The Rockefeller Brothers Fund, The Florence and John Schumann Foundation, The Tides Foundation, and individual donors.

About Defenders of Wildlife

DEFENDERS OF WILDLIFE is the only national membership group that devotes all its resources to advocacy for the protection and restoration of all species of wild animals and plants in their natural communities.

Founded in 1947 and headquartered in Washington, D.C., Defenders and its 80,000 members and supporters have consistently worked, through education and advocacy, to protect and enhance wildlife species and their habitats.

At the forefront of wildlife conservation campaigns, Defenders' staff, regional representatives, and volunteer activists throughout the country rely on public education, litigation, scientific conferences, expert testimony, and grass-roots action to accomplish their many goals and objectives.

Landscape
Linkages and
Biodiversity

Landscape Linkages and Biodiversity

DEFENDERS OF WILDLIFE

Edited by Wendy E. Hudson

ISLAND PRESS

Washington, D.C. □ *Covelo, California*

QH
75
. L28
1991

Library of Congress Cataloging-in-Publication Data

Landscape linkages and biodiversity / edited by Wendy E. Hudson.
    p.   cm.
   Includes bibliographical references and index.
   ISBN 1-55963-108-2.—ISBN 1-55963-109-0 (pbk.)
   1. Biological diversity conservation.   2. Habitat conservation.
I. Hudson, Wendy E.
QH75.L28   1991
333.9516—dc20   91–15607
                                    CIP

Contents

Foreword

THE NATION IS experiencing a serious crisis in the conservation of its biological diversity. Our single-species tactics and fragmentary approaches to protecting wildlife and wildlife habitat simply are not working.

While the Endangered Species Act has had some notable successes, it is not keeping up with the nation's biodiversity crisis. A recent Interior Department internal audit notes that thirty-four animal and plant species have been determined to be extinct without ever having received full benefit of protection provided by the Endangered Species Act. And while about 550 species have been listed since the act became law, another 600 known and 3,000 probable threatened and endangered species have never been listed.

Individual species protection is still critical, but it is essentially useless without an effort to protect habitat as well. Habitat loss, fragmentation, and isolation are the most common causes of species extirpation. As natural areas continue to be disrupted by human activities, animal and plant populations become isolated in "island habitats" where genetic inbreeding, depredation of large species, and proliferation and domination of human-adapted species all interact to increase rates of extinction.

Clearly, the growing list of threatened and endangered species and the alarming rate of habitat loss indicate that more needs to be done at an earlier stage. Early intervention will help prevent species from becoming threatened and endangered and habitats from becoming fragmented and degraded. What we need is a national biodiversity network to maintain the integrity of natural ecosystems and their living organisms.

The first step in establishing a national biodiversity network is to develop biological inventories to identify ecologically significant lands that are not yet protected. Known as "gap analyses," these pioneering inventories are providing valuable information on a large scale about ecologically significant lands that are not yet protected. Use of the gap analysis will help form the basis of a comprehensive habitat protection system. With data from the gap analysis in hand, the next step is to protect the habitats identified as being ecologically significant. These two steps lay the foundation for a national biodiversity network.

Defenders of Wildlife has raised the banner of biodiversity conservation as its central mission for the nineties. In 1990, at the fifty-fifth annual North American Wildlife and Natural Resources Conference, Defenders convened prominent scientists and conservation advocates from around the nation to discuss the many strategies for implementing biodiversity conservation efforts. Credit for the idea behind the symposium goes to Sara Vickerman, Defenders' Oregon-based regional program director, who has been instrumental in promoting a wide array of biodiversity issues. Wendy Hudson, communications coordinator, organized the symposium and devoted long hours to editing the manuscript of this book.

The essays from this symposium and the edited transcripts of the audience/panel exchange largely comprise the contents of this book. In a refreshingly nontechnical manner, the authors collectively examine the theoretical underpinnings and practical applications of biodiversity conservation and landscape linkages. Overviews from outside experts—those who did not participate in the meeting—add a fresh dimension to the book. We hope that this collection of essays, overviews, and transcribed discussions will remove some of the confusion surrounding the concept of biodiversity conservation and inspire a continuing dialogue on this critical issue.

Defenders' mission for the nineties is ambitious, but there is none more important. Time is running out. Development pressures everywhere are mounting, and wildlands and wild species are disappearing at an alarming rate. The nation urgently needs a fresh, integrated approach to protecting our biological heritage. If we are to have any hope of preventing countless plant and animal species from becoming extinct, our conservation efforts must focus on protecting diverse habitats—on protecting this nation's biological diversity.

—M. RUPERT CUTLER
  Past President, Defenders of Wildlife
  Currently director, Lewis and Clark Environmental Education
  Center, The River Foundation, Roanoke, Virginia

Editor's Preface

ON MARCH 17, 1990, Defenders of Wildlife initiated its "Mission for the Nineties" by sponsoring a half-day symposium in Denver on the issue of biodiversity conservation. Held in conjunction with the fifty-fifth annual North American Wildlife and Natural Resources Conference, the symposium was divided into two sessions, each with a panel of specialists who delivered brief talks and fielded questions from the audience.

The panel debated a wide range of issues in an effort to bring the concepts of biodiversity conservation closer to the several hundred attending specialists in wildlife research and management. For the entire afternoon, a standing-room-only crowd gathered in the large hall, and by day's end it was apparent to all that many valuable ideas had been raised. Far from conclusive, the symposium left many questions unanswered and sparked many conflicting passions. But it was precisely these ambiguities and differing opinions that pointed up the complexity of biodiversity conservation—the need for further research, for a broader exchange of ideas, for more cooperative partnerships, and, above all, for perseverance.

We at Defenders felt the symposium raised such valuable ideas that they deserved to reach a wider audience in the form of a book. Because the essays touched on such a wide range of biodiversity conservation issues, the challenge was to organize the book with sufficient continuity and flow. Accordingly, the essays were grouped into three recurrent themes of the symposium: Conserving Biodiversity, Countering Habitat Fragmentation, and Reintegrating Humans and Nature.

We also felt that adding an overview for each of the three sections would lend further continuity and flow to the book. We selected three reviewers—Don Waller of the University of Wisconsin; Blair Csuti of the University of Idaho; and Allen Cooperrider, formerly with the Bureau of Land Management—on the basis of their professional stand-

ing within the biodiversity conservation community and their contributions to the field. We asked them not only to provide a brief overview for their section but also to critique the essays and proffer their
own opinions.

Douglas Chadwick, a Montana biologist and writer, attended the
symposium and later produced a special report, "The Biodiversity
Challenge," which appeared in the May/June 1990 issue of *Defenders
Magazine*. To round out the book, we excerpted this highly acclaimed
report for use as the Introduction. Also helping to round out the book
are the edited transcripts of the audience discussion that appear at the
end of each section. Less formal than the essays, the discussions reflect
the real-life dilemmas and frustrations of day-to-day conservation.
Certainly the responses of the panelists confirm the complexity of the
discipline, and the juxtaposition of the audience discussion with the
essays brings home the difficulty of converting theory to practice.

Defenders of Wildlife offers this book as the start of a continuing
dialogue on the ever expanding subject of biodiversity conservation.
As habitat loss, fragmentation, and isolation become more ubiquitous,
the need for identifying and protecting large, ecologically diverse
lands becomes all the more pressing. Defenders of Wildlife hopes that
by sponsoring such symposia and publishing the fertile ideas and
controversies of the discipline, it will contribute to making biodiversity conservation a reality in America.

—W. E. H.

Acknowledgments

DEFENDERS OF WILDLIFE gratefully acknowledges the financial support received from the following organizations to help cover the costs of the symposium upon which this book is based:

Breedlove, Dennis and Associates of Winter Park, Florida
Fanwood Foundation/West, Montana
The Simpson Trust of the Rockefeller Foundation, New York

Introduction

SINCE THE SETTLEMENT of North America began, more than 500 species and subspecies of native plants and animals have become extinct. Seven listed species have vanished since Congress enacted the Endangered Species Act in 1973. Several others, among them the California condor and the black-footed ferret, have become extinct in the wild but survive in captivity. The threatened and endangered lists have grown to include 565 species. Nearly half of them lack any recovery plan. Meanwhile, close to 4,000 additional United States species are considered candidates for listing. Two or three hundred may already be extinct, and that is probably a bare minimum. Some say the true number of candidates is closer to 6,000. A recent study suggests that California alone may have 220 animal and 600 plant species threatened with serious reduction, if not outright oblivion.

Clearly, the number of lifeforms at risk on this continent is continuing to increase and is doing so at an accelerating rate. Chart the upward curve of that rate and it becomes clearer still that we are on the verge of massive biological impoverishment, nearly on a par with that forecast for developing nations in the tropics. The truly frightening thing—and the reason we need to reexamine our whole conservation agenda in this country—is that we ended up here on the brink so quickly despite the Endangered Species Act and despite having what many would consider exemplary nationwide systems of parks, wilderness areas, wildlife refuges, and other preserves. What are we doing wrong?

To begin with, in the view of several of the workshop panelists, we have been too preoccupied with trying to save species, one at a time, after they are already in trouble. Conservation biologists liken this strategy to doling out emergency-room treatment on a case-by-case basis in a sort of popularity contest. Most of our attention has gone to a relative handful of "charismatic" creatures, mostly birds and mammals with qualities of majesty, intelligence, or cuddliness that we can easily relate to.

Save the grizzly! Save the sea otter! Save the whooping crane! But what about the small, the slithery, and the leafy—the less conspicuous species that make up the vast majority of our biota and the majority of its candidates for the endangered list? These species get little publicity or support. Yet they are equally vital to the stability of ecosystems, and their promise as potential sources of food, fiber, or pharmaceutical products matches that of large or showy forms.

Even if our emergency-ward efforts become less arbitrary, and even if time, money, and personnel were not in such short supply, we could not hope to get ahead of the curve as long as we are only reacting to crises. The way out is to begin practicing more preventive care. That means doing a better job of safeguarding healthy flora and fauna at the biological-community level. In particular, it means tending to those communities with the greatest variety or richness of native species. Focusing on such areas of high biological diversity—biodiversity for short—gives us the best chance of keeping the most species from becoming endangered.

In the opinion of Michael Scott, a research biologist with the U.S. Fish and Wildlife Service in Idaho, "The bottom line in conservation biology is not how many species we save from extinction in the next decade but how many species will survive the next hundred years or more."

"If we could identify all the natural communities present in this country at the time of European settlement and keep examples of each intact, we would have about 85 percent of all our native species conserved," estimated panelist Ben Brown, Colorado program director for The Nature Conservancy. "Perhaps what is needed," offered Felice Pace of the Klamath Forest Alliance, a California citizens' group, "is not an Endangered Species Act but an Endangered Ecosystems Act."

Perhaps so. The primary reason so many creatures are in trouble is that much of their habitat has been lost and what remains is being badly fragmented. For many species, even the largest fragments are proving too small and isolated to sustain them over the long run. To understand why, we need to turn to the realm of ecological theory known as island biogeography. If nothing else, it explains why it is becoming difficult to talk with conservationists these days without hearing about islands and bridges.

It only makes sense that large geographical areas should hold more animals than small ones. But why will the larger areas also hold more kinds of animals? Scientists first recognized a correlation between species richness and area early in this century. Then, during the 1960s, ecologists studying the biota of islands developed a set of principles relating the size and isolation of areas to the number of species the areas can support over a period of time. The findings most relevant to conservation as a whole come from the study of what are called land-bridge islands that used to be attached to the mainland before being cut off by the sea.

The longer an island has been isolated, the less its flora and fauna will have in common with the communities on the mainland. Many of the island's original species will have gone extinct, while many of the survivors will have evolved into uniquely adapted forms. Over time, the biota of the smaller islands becomes especially skewed with a marked drop in overall biodiversity. Such areas simply cannot hold enough members of certain species, especially the larger animals, to maintain a stable gene pool. Small, insular populations lack the genetic flexibility to cope with changes in the environment, and their vulnerability worsens as undesirable traits accumulate through inbreeding. Sooner or later the result is extinction. The loss of each species ripples through the community, further destabilizing the balance among survivors and often triggering more extinctions.

If you compare a particular island to one that is generally similar but only a tenth as large, the smaller island may be expected to hold only about half as many species—and often far fewer. Now turn to a habitat slated for development on the mainland. Suppose a tenth of the original area is to be set aside in a preserve. That may be a fairly generous percentage in many people's minds, especially considering that just 5 percent of the lower forty-eight states as a whole lies within protected areas. They might assume that this preserve—this island within a sea of disturbance—ought to be more than enough to maintain a representative sample of all the community's original inhabitants. They might be right for a few years. Then every year afterward they would be more and more wrong.

Not surprisingly, the state with the highest percentage of imperiled species is Hawaii, our one island state. The implications of island biogeography for fragmented habitats and ecosystems in the other forty-nine states are obvious. This is not to say that the parallels are exact. After all, habitat "islands" on the mainland are surrounded by other types of landscape—not by water—and so there is a greater chance that a declining population with a shrinking gene pool can be recharged by a few individuals making their way into the area. By the same token, however, predators or competitors from nearby habitats may invade the "island," where certain small populations are held hostage, and bring about local extinctions even faster than might be expected.

A remnant of pristine forest harboring rare plant species, for example, may be overgrazed and irrevocably altered by the deer thriving in nearby logged forests and farm country. Similarly, a number of refuges with perfectly good wetland habitat have nevertheless become poor producers of waterfowl and other aquatic birds because so many eggs and young are taken by skunks and raccoons. Like deer, these adaptable, omnivorous predators prosper in many habitats altered by human activities. Their populations can reach unusually high levels, since humans have also eliminated the larger predators that once held skunks and raccoons in check.

The main lesson of island biogeography is this: We cannot tuck species away in little preserves, as if we were storing pieces in a museum, and then come back a century later and expect to find them all still there. The essence of life is change. Organisms are constantly growing, interacting, adapting, evolving. Their numbers and distribution across the landscape fluctuate in cycles linked to climatic patterns and to other, less understood rhythms. They are defined as much by their place in food webs and nutrient flows as by their own physical traits or any current geographic location. Many alter their range and behavior under different conditions. Some assume entirely new behavior through learning. In short, an ecosystem is not a collection of plants and animals. It is a seamless swirl of communities and processes. If you don't save the processes, you won't save the parts. So if you're going to create a preserve, you had better make it a big one.

In a provocative article published in 1987, ecologist William Newmark demonstrated strong similarities between extinctions on true islands and the disappearance of forty-two types of native mammals from the fauna of fourteen North American parks over the past several decades. The badger and black bear slipped away at Zion; red fox, northern flying squirrel, beaver, and pronghorn at Bryce Canyon; gray fox, spotted skunk, ermine, mink, and river otter at Crater Lake; and so on. All these species vanished—even though they were present when the parklands were established and they were protected thereafter from direct harm at the hands of humans as well as from indirect harm through development.

The plain fact is that most of our existing preserves have the same problems as fragments of habitat elsewhere across the nation: They are too small and isolated to guarantee the long-term survival of many of their wild residents. Few, if any, of our reserves are big enough to sustain species with very large home-range requirements, such as the wolf or grizzly; nor are they in sufficient numbers to ensure an adequate gene pool century after century. Evidence exists that the average size of Yellowstone's isolated population of grizzlies is becoming smaller, possibly as a result of inbreeding. The population holds barely forty adult females.

"In all of Florida, we have only one or two reserves big enough for a single Florida panther," panelist Larry Harris of the University of Florida's department of forest resources and conservation informed the Denver audience. "And 85 percent of the new preserves we have acquired are too small to encompass the home range of a single otter." Nor are more than a handful of preserves in the nation big enough to withstand the impacts of hurricanes, fires, floods, and other catastrophes. We call them catastrophes, anyway, seeing them as rare and destructive. But in nature's long-term perspective, they are regular events and part of the cycle of habitat succession. Healthy ecosystems can absorb them. But precious few preserves have room enough for their inhabitants to respond to even minor climatic changes—cycles of drought, for instance—without losing diversity. (How will they cope with global warming?) And no preserve is big enough to protect all the migratory species that use it.

Many other preserves, whether big or small, are in the wrong place to begin with. Some were set aside chiefly on the basis of scenery, recreation, or the needs of a few favored game animals, others because no one had much use for them at the time. Often they have little do with ecological boundaries or conserving biodiversity.

Panelist Reed Noss, at the time a biodiversity specialist with the Environmental Protection Agency Research Laboratory in Oregon, has published a paper pointing out that of 261 ecosystem types in the nation identified by another study, 104 (or 40 percent) were not represented at all in our national wilderness preservation system. Wilderness areas are supposed to preserve samples of primeval America. They do. But too often they exemplify land that didn't have enough timber, fertile soil, potential housing sites, or other commercially valuable resources to block passage of a wilderness proposal. Examples of the incalculable wealth of life that the New World once held are harder to come by.

An analysis in California revealed that 95 percent of the state's various alpine habitat types were present in existing reserves. This is because the bulk of California wilderness areas are stuck up on mountain ranges, safeguarding chiefly rocks and ice. The future looks bright for high-altitude lichens. Meanwhile, less than 1 percent of California's riparian, or streamside, habitat types—the communities with the greatest species richness—are protected. The situation is even starker in more arid states such as Arizona and New Mexico. There, riparian zones comprise only 1 or 2 percent of the total land area to begin with, and 90 percent of these habitats have been lost or degraded over the last century. Just a tiny fraction of what is left is protected. Yet up to 80 percent of the native vertebrate species in the region depend upon riparian habitats during at least some part of their life cycles. Of our mighty sweep of savanna—the Great Plains that dominated the interior of the continent—we have preserved almost nothing, either of the original prairies or of the thundering biomass of hooved animals they once supported.

Nationwide, the areas administered by our four largest federal land-managing agencies—the National Park Service, Forest Service, Fish and Wildlife Service, and Bureau of Land Management (BLM)—add up to a vast expanse. But when a study team looked at this acreage in blocks of 10,000 hectares or more, they saw that 22 percent of the various kinds of ecosystems recognized in this country are not represented. Another 29 percent are underrepresented. That leaves just under half in blocks of adequate size. And the majority of this acreage is not in preserves but on Forest Service and BLM land open to multiple use, which means everything from clearcut logging to open-pit mining.

The remedy? First, identify areas with the highest levels of biodiversity. For years, The Nature Conservancy has been working to inventory communities and sensitive species through its Natural Heritage database program, and Mike Scott's "gap analysis" program has been building on this database. Using powerful computers and a sophisticated digital program, the gap analysis technique identifies areas of species richness and shows where gaps occur in the network for protecting species richness.

The second step, after identifying areas with the highest levels of biodiversity, is to protect as many as possible in the largest blocks possible. Since, realistically, political and economic interests will make it difficult in many cases to create new preserves, we will have to patch blocks together by selectively adding to the old ones. We should also buffer them as best we can with zones that restrict at least the most destructive kinds of development and pollution in critical habitats along the boundaries of fully protected lands.

Panelist Gary Barrett of the zoology department at Miami University in Ohio spoke of the potential for blending agricultural practices with the workings of neighboring ecosystems, which he defined in a fresh way. "Ecosystems are not something for nature lovers," he said. "They are solar-powered systems of growth with integrated pest management; unsubsidized, self-sustaining, cheap to run, and highly productive." The most difficult and expensive way to pursue conservation would be to rebuild biological communities practically from scratch. Yet this emerging field of conservation, called reconstruction biology, is what may be required in some heavily populated and developed regions of the East.

The final step, after identifying areas of high biodiversity and protecting as many as possible, is to make connections. Because the protected areas we fashion will never be big enough for some species no matter what their final shape, we must begin linking them with corridors for the movement of animals and the dispersal of plant seeds. With such land bridges, we can transform an archipelago of lonely habitat islands into a functional unit grand enough to preserve the integrity of an entire ecosystem.

Landscape linkages, land bridges, wildlife corridors, greenways, shelterbelts, turkey trots—call them what you will. They can be as big as the 34,000 acres recently purchased by the USDA Forest Service and The Nature Conservancy in Pinhook Swamp to link the 160,000-acre Osceola National Forest in Florida with Okefenokee National Wildlife Refuge in Georgia, the largest wildlife refuge in the East. Or they can

be as small as a hedgerow, long the backbone of many a small crea-
ture's habitat in Europe.

Panelist Michael Soulé of the Board of Environmental Studies at the
University of California, Santa Cruz, spoke of the pitfalls as well as
the promise of wildlife corridors. Certain kinds could become avenues
for the spread of exotic or "pest" species; others in the wrong location
could lead to mingling of communities that would normally remain
separate and distinct.

As with the underpasses recently constructed by citizens of Am-
herst, Massachusetts, to enable salamanders to cross a busy road on
their way to and from seasonal breeding ponds, there are endless
opportunities for creative solutions. Jogging and riding paths can serve
as corridors for small mammals and birds in suburban areas. The
Nature Conservancy has acquired what panelist Ben Brown called
"our Amtrack reserve," more than 100 miles of railroad right-of-way
along an abandoned set of tracks in the Upper Midwest. "Together
these linkages hold more promise," said Keith Hay of The Conservation Fund, "for the conservation of the diversity of life than any other
management factor except stabilization of the human population."
"We don't even have to say we want corridors," commented Felice
Pace. "We want continuity of life."

Of course, it's not going to be that easy. But then again, it doesn't
have to be all that hard once our public agencies "embrace biodiversity
on an equal basis with their other goals," as panelist Hal Salwasser,
deputy director of wildlife and fisheries for the USDA Forest Service,
said after the workshop. Salwasser continued: "To put it in simplest
terms, the ultimate goal is to have all the native species in the natural
communities we inherited able to play out their evolutionary dramas.
And that means they've got to be around a long time. Therefore, the
landscape must be such that these animals—and plants—can move
around. We need to be looking at big chunks of country. We need
enough of the right kinds of habitat in the right configurations. To get
that, we are also going to need unprecedented levels of cooperation
among agencies and between those agencies and the private sector."

That's one of the things I like best about the concepts of biodiversity and landscape linkages. Boil down all the scientific formulas
and jargon, and you're talking about learning how to get along with
your neighbors. Fragmentation of habitat in this nation has its
counterpart—and, I think, part of its cause—in the fragmentation of

resource management. Responsibilities are divided among a welter of competing agencies and organizations, each with a different set of goals.

Some 80 percent of our national park boundaries adjoin national forest land. While Park Service rangers might be reprimanding a camper for picking a wildflower, Forest Service officials may be supporting oil and gas development just across the border in wildlife habitat of the same or better quality. Even within the same agency, different divisions may work at odds with one another. In segments of our national wildlife refuge system, conservation of wildlife diversity is fairly far down the list of priorities, coming after waterfowl hunting, fishing, powerboating, other recreation, grazing, and mining.

My point is less to criticize than to make it plain that we have no common guidelines among the different federal resource agencies. This is because we have no macrostrategy for conservation in the United States. For lack of a coherent national policy, we have no detailed national inventory of biodiversity. And for want of a detailed inventory, we have no way of planning to make certain we maintain at least 90 percent, 75 percent, even 50 percent of our native biological communities. No single agency manages enough land or has enough personnel to carry out the job alone, even if it has a clear directive to do so. The agencies don't even share a standard method for identifying habitats. Each goes by a somewhat different scheme tailored to its own purposes. As a result, one agency might recognize twenty-seven vegetation types in an area but another could come up with sixty-two. We plainly need to work for linkages among resource managers as well as within ecosystems.

Happily, that sort of connectivity does seem to be coming about. Several panelists mentioned the Greater Yellowstone Ecosystem as a prime example. There representatives of three state wildlife departments, several state forests, five national forests, BLM, the Park Service, and the Fish and Wildlife Service have formed an interagency committee. What drew this group together at the outset was the problem of saving grizzlies, which ignore human boundaries while occupying home ranges of up to 1,000 square miles (in the case of adult males). Once the dialogue began, however, the managers found themselves dealing with the welfare of elk, trumpeter swans, and an array of other animals that have always treated the greater Yellowstone ecosystem the way the agencies are beginning to: as common ground.

At the same time, an unusually wide assortment of private interests—from backpackers and birdwatchers to hunting guides and hotel owners—formed the Greater Yellowstone Coalition. They, too, favor a holistic approach to protecting the ecosystem. The interagency committee and the coalition are working together closely. Similar collaboration has begun in and around the largest park in the East, the Great Smoky Mountains in North Carolina and Tennessee. And, less than a week after the Denver workshop, the BLM announced it would begin working closely with The Nature Conservancy to improve protection of wildlife communities on all lands the BLM manages. Here is a joint effort of great importance, since the BLM oversees 270 million acres of the nation's public lands.

Reed Noss has written: "Biodiversity is no frill. It is life, and all that sustains life. It is worthy of respect. Maintenance of biodiversity must become our primary mission as a society, the principle that guides all resource use." That statement is hard to argue with. Simplified ecosystems, whether monocultures such as crop fields and pine plantations or habitat islands with impoverished biota, are subject to sudden collapse from a minor shift in their environment. It is an ecological truism that in diversity lies stability. And in the stability of ecosystems lies the stability of the cultures that depend upon them. Their richness is ours.

One may well ask why it has taken so long for the concepts of biodiversity and landscape linkages to begin making headway in this country. These concepts have been applied to developing nations for over a decade. I guess it took a barrage of headlines about tropical deforestation, desertification and famine in sub-Saharan Africa, decimation of fisheries in the once supposedly boundless seas, acid rain drifting across international borders, and a dozen other planetary disasters to make us realize that there really can be an end to nature, an end to evolution in the form that we have known it.

Loss of the tropical forests may be speeding up global warming. It is surely decimating the richest trove of species on the globe. Why, then, are we cutting down original forests in such places as Florida and the Pacific Northwest at a rate two to three times that in the Amazon? We fancy ourselves to be a developed country, not a developing one. We don't need a nationwide conservation plan, we say—that's for Third World countries reaching an ecological crisis point. We have dozens of agencies already in place for managing game and forests and hiking areas and. . . . What *are* all these species in our own country doing tumbling by the score onto the endangered list?

Larry Harris offered another thought on why we are only now taking biodiversity to heart: "It's partly because we swept across this continent so quickly, exterminating bison and wolves and so much of our big wildlife in the contiguous states that we never really knew what was here. We destroyed the great patterns of migration and interaction. We destroyed the communities essential to our vision of what animals need. By the time we got around to reestablishing any, they were nearly all in tiny, isolated fragments, and that's how we went about managing them."

As it has become apparent that traditional wildlife management is not sufficient to protect the full variety of species, conservationists have begun to push for more attention to "nongame" species. Nongame biologists have been added to many wildlife department staffs. But the programs, for the most part, have remained small and inadequately funded—poor second cousins to game management. Isn't it odd that a vaguely negative term like "nongame" should be used to describe the overwhelming majority of our native fauna, as if they were somehow nonessential or at least less essential? It's as curious as the "nonconsumptive user" to describe the overwhelming majority of people who enjoy going out to see wildlife. Only about 9 percent of Americans hunt game, whereas the numbers involved in birdwatching, amateur naturalist studies, and other activities related to wildlife increased from 55 percent in 1980 to 74 percent by 1985. They are higher yet today. Here is another reason for the growing interest in biodiversity and linkages. The public wants wildlife, all kinds of wildlife. People want to watch it, to learn more about it, and simply to know that it's there. They also want to know why it's *not* there.

"Keep in mind that biodiversity is still a lot more talk than action in this country. It is certainly a lot more talk than solid programs and funding as far as the agencies are concerned," cautioned Michael Soulé. "If that doesn't change fairly fast," he added, "we won't be doing any real conservation biology twenty years from now. It will all be restoration biology."

Let's hope not. The conditions are right for conservation to evolve again—this time to go beyond endangered species and beyond nongame. Our challenge now is to conserve the very nature of nature, which is the power to connect, to sustain, to heal, and to invent; to keep filling the world with an infinite variety of wonders. That is how it's supposed to be. That is how it can be if we move beyond saving bits and pieces of nature and plan for the preservation of wholeness. The question is how to spread the word among enough citizens and policymakers while there is still time. That is what biodiversity and landscape linkages are all about.

—DOUGLAS H. CHADWICK

I

Conserving Biodiversity: A Unified Approach

Introduction

DONALD M. WALLER

FOR CENTURIES, SOCIETIES have assumed that the natural world constitutes "resources" that we are both free and obliged to use. We have also assumed, collectively, that the many and cumulative transformations of the landscape for agriculture, forestry, and other human purposes would not threaten its sustained utility or the survival of its original biota. These comforting myths were difficult to shed as long as there was new land for colonization and the denaturation of our landscapes occurred slowly enough to go unnoticed. Accelerated population growth, industrialization, and other "development" in the late nineteenth and twentieth centuries, however, have shaken our complacency by forcing us to confront the many chronic and pervasive side effects of our extractive cultures. The shock of recognition occurred first in response to the immediate and dramatic threats of foul air, dirty water, and toxic wastes. More recently, the public has begun to recognize the more insidious disruptions occurring in our biotic environment: As natural habitats dwindle, rare species go extinct while weedy species proliferate.

Historically, the motivation for conservation was the desire to sustain the land's ability to produce resource commodities. The management of public forests, rivers, parks, and rangelands was entrusted to professionals trained in game management, engineering, and forestry. These professionals naturally centered their attention on ensuring abundant populations of the species used directly by humans. Growing concern about declines in natural populations of rare and threatened species in the 1960s culminated in the passage of the Endangered Species Act of 1972. This act for the first time mandated measures to identify and restore populations of threatened and endangered species

3

and focused federal attention (and even some funds) on the hazards these species face. It was natural under such a law to transfer the species-by-species approach common in game management to threatened and endangered species using techniques like captive breeding, active reintroductions, and intensive habitat management. These rescue efforts succeeded in cases like the peregrine falcon where specific threats (such as pesticides) could be identified and addressed. Because these programs are elaborate and expensive, however, conspicuous furry or feathery species have been the primary beneficiaries. Furthermore, even such conscientious efforts fail when too little appropriate habitat remains to allow successful reintroduction or recovery.

These experiences, as well as the scope and subtlety of assaults on biotic diversity, have led many conservationists to propose more inclusive approaches to conserve diversity. There is increasing recognition that the Endangered Species Act has failed as an instrument to protect the broad spectrum of species, communities, and habitats threatened with serious declines and accelerating extirpation. While it is convenient to attribute this failure to the lack of adequate funding and consequent slow rate of listing species and implementing recovery plans, many now question whether a species-by-species approach is really the most effective means to protect biodiversity. MacArthur and Wilson's (1967) theory of island biogeography with its explication of species/area relationships and "relaxation time" has provided a rich source of general insights into species diversity. This theory has also focused attention on key questions regarding the size and shape of areas needed to maintain diversity—culminating in the rancorous, if short-lived, "SLOSS" (Single Large or Several Small) debate (Soulé and Simberloff 1986).

The workshop sponsored by the Defenders of Wildlife was designed to address the shortcomings of previous approaches to conservation and review recent scientific and administrative initiatives to remedy them. In this part of the book, federal agency scientists associated with the practical problems of conservation explore new strategies to stem the tide of species losses. These authors present a wide range of alternative approaches, but all are concerned with the holistic issues of geographic scale, community and landscape dynamics, and interactions between conservation ideals and political realities.

Although a general consensus has emerged within conservation biology regarding the need to establish large and connected networks of reserves, we have yet to agree on the best means to achieve that end. Here I review each chapter's contributions to this discussion. I also raise issues that may have been understated or overlooked in the symposium. The diversity of ideas presented and the open debate surrounding their implementation reflect both the vitality of this young field and its struggles to translate scientific ideals into effective conservation programs.

DIAGNOSIS

All the authors in Part I emphasize the need to conserve large areas and coordinate conservation efforts across regions. The first ingredient in an effective conservation project is, of course, to follow Aldo Leopold's admonition to save all the pieces. Chapter 1 by J. Michael Scott, Blair Csuti, and Steven Caicco directly addresses this primary need by reviewing a new technique they have developed: "gap analysis." Their approach is based on maps of vegetation types that can be used to predict the range and abundance of native terrestrial vertebrate species. This surrogate indicator of biodiversity is designed to identify those habitat types not currently well protected within national parks, wilderness areas, or diversity management areas. Coupled with knowledge regarding the rate of loss or conversion of those habitat types in nonconserved areas, such an approach provides a powerful tool for assessing the threats to diversity on a regional scale.

Gap analysis allows us to quickly identify particular habitats that lack protection—thus allowing public and private conservation agencies to move decisively and in unison toward protection. It provides a single technique that can be duplicated across many regions to systematically assess degrees of habitat protection over large areas. The authors recognize this potential and urge that their technique be adopted widely, particularly in the West. By using the association between animal populations and vegetation type, gap analysis circumvents the need for exhaustive and expensive biological surveys. Advances in community classification, remote sensing, and geographical data analysis have all facilitated the development of this integrated approach.

Given these obvious advantages, gap analysis deserves to be widely adopted and implemented. In pursuing this approach, however, its

limitations should also be kept in mind. First, certain technical difficulties may arise—for example, there may be inconsistencies between
the spatial scale of imaging resolution and the biological scale of
vegetation maps or range maps for threatened species. Furthermore,
the quality and quantity of biological information vary considerably
from place to place, making assumptions regarding the correspon-
dence of habitats and populations, even of vertebrates, risky. Moreover, the polygons of each habitat type within the database do not
portray the successional status of those habitats and also ignore microhabitat information critical to predicting associations with certain
species. The degree to which such imprecision limits the utility of gap
analysis awaits further study. Thus we are confronted with an ironic
situation: Validation of gap analysis will require precisely the up-to-
date and accurate biological inventories that the technique is designed
to sidestep.

There is a more serious concern, however. Even with completely
accurate inventory data and reliable analysis, we have much to learn
regarding the minimum area and habitat requirements of rare and
threatened species. Although deleterious edge effects and habitat fragmentation have emerged as critical topics in conservation biology, gap
analysis has yet to incorporate these effects adequately. Many top carnivores are generalists that use a variety of habitat types, yet they may be
very vulnerable to geographic features in their environment such as the
proximity or connectedness of protected areas (see Chapter 2 by Noss).
Thus, inferences based on simple tallies of the total area within each
type of habitat could be seriously misleading. Moreover, discussions of
"scale of protection" can be disturbing because of a naive tendency to
assume that national parks provide the most protection, wilderness
areas second best protection, and private land no protection for rare and
threatened species. Experience has taught us the dangers of assuming
that parks afford automatic protection (Janzen 1983; Chase 1986), and
rare plants and animals often find refuge on private lands.

Some of these difficulties are technical in nature and could be addressed via modifications of their approach. For example, information
regarding the actual sizes of protected and unprotected habitats and
their spatial proximity could be included within the database, at least
in principle. Algorithms incorporating such effects could then be
developed. The potential for such improvements, however, clearly
rests on the presence of detailed field studies documenting the exis-

tence and scale of such effects for all species of concern. Even sophisticated prediction schemes like gap analysis should not be accepted as substitutes for the mundane but concrete data derived from extensive biological surveys. There is already an alarming tendency within federal agencies to rely too heavily on predicted occurrences of the biota—leading us to the awkward situation that even national parks lack systematic monitoring of their large mammals (Newmark 1987).

Thus, gap analysis should be viewed as an important but preliminary tool for monitoring trends in land use and identifying types of habitat that should be protected under biodiversity management. Given its utility and appeal, it will probably be widely applied and extended in the years to come. Like any good tool, its usefulness will depend on the skill of its users. Conservation biologists should learn how to use it carefully and seek to extend it in order to incorporate more geographical complexity and known dependencies of diversity on minimum areas and connectivity of habitats.

REMEDIES

Of course, even improved forms of gap analysis can provide only a preliminary diagnosis of conservation ills. To restore landscapes to full biological health will require coordinated treatment at all spatial scales. Some of the most interesting research in conservation biology has concentrated on questions regarding scale. The focus on individual species and particular sites has broadened to include concerns over "greater ecosystems," "minimum dynamic area" (Pickett and Thompson 1978), natural rhythms of disturbance (Petraitis et al. 1989), and the need to protect species "wholesale" through habitat conservation. The immediate habitat needs of large, widely roaming carnivores as well as the more diffuse relationships between population size, extinction probabilities, and species diversity all support the idea that large, well-planned, and biologically connected networks of reserves are needed to conserve biodiversity (Solheim et al. 1987).

Reed Noss in Chapter 2 champions the importance of geographic scale and "landscape linkages" in planning networks of conserved lands. He argues in particular that conservation efforts should begin to focus on maintaining "connectivity" among natural habitats. While emphasizing the obvious need to conserve core areas of adequate size, he contends that wide corridors comprising a variety of habitats could often remedy (or, better, prevent) increased habitat fragmentation. We catch glimpses of the genesis of these ideas as he recounts his efforts to design an ambitious network of reserved and semiprotected areas across Florida to protect the Florida panther and other wide-ranging species. Such schemes are based on the growing realization of many conservation biologists that we must pay more attention to sustaining dynamic population, community, and ecosystem processes if we are to conserve diversity. Only conserved areas of adequate size and connectivity maintain the opportunities for genetic exchange, natural disturbance dynamics, and immediate local recolonization ("rescue from within") that can maintain diversity over the long term.

Noss draws on a growing body of research, many specific examples, and his own experience to argue convincingly for the need to maintain continuous habitats within and among regions. In proposing specific networks for conservation, he displays sensitivity to the biological specifics of different situations. He wisely resists the temptation to found a new fad school of "corridor" conservation based on strict formulas or simplistic prescriptions. Instead, his chapter leads by example to show us how we need to work at a variety of spatial scales, using the best of our talents and knowledge, to design biologically functional networks of reserves. An excellent example of this sophistication lies with Noss's suggestion that we *disconnect* certain habitats. As we promote linkages between natural areas, Noss argues, we should simultaneously restrict connectivity between artificial habitats to hinder the spread of weedy species and exotic pests. Achieving both goals will not be easy. But it is essential if we are to resist biotic impoverishment through homogenization of our landscapes.

PUBLIC HEALTH

In contrast to considerable advances in conservation biology, the training of new scientists, and successes with particular species, we continue to lose species and habitats at an accelerating pace. These failures have convinced many conservation biologists of the need for a more politically unified approach. Conservationists now agree on the need to systematically design networks of large core areas—connected when possible and managed using techniques that mimic original patterns and scales of disturbance. They also agree on the need to guide these efforts using the best possible science—injected at the highest levels of planning and implemented through policies coordinated among the various levels of government charged with land management (together with private organizations like The Nature Conservancy).

In contrast to the unanimity of opinion regarding prescriptions to maintain diversity, conservationists diverge widely on the issue of how to attain these goals. Some, content to work within existing structures, believe that ad hoc administrative agreements provide adequate structure and flexibility for dealing with particular situations. Others point out that the United States currently lacks any umbrella legislation regarding diversity, relying instead on a small and disconnected set of federal and state acts and regulations. These limited responsibilities, fragmented among several government agencies, prevent strong federal leadership, systematic planning, and sustained actions to conserve diversity.

Allen Cooperrider of the Bureau of Land Management begins Chapter 3 by reviewing the federal government's expanding spheres of interest in conservation and the concomitant extension of conservation efforts to an ever wider set of species and habitats. He focuses his discussion on the multiple threats faced by western rangelands—not only the direct effects of grazing, such as soil erosion, water diversion, and the invasion of exotics, but also a variety of unanticipated indirect effects, such as declines in certain bird species and competitive effects on desert tortoises. These effects are widespread throughout western lands, yet often they go unnoticed by a public that has seen little else and may therefore fail to appreciate how sensitive such arid lands are to disturbance. Ironically, it was exactly this landscape that spawned Aldo Leopold's formative ideas regarding land health, plant/grazer interactions, and the need for wilderness areas. Cooperrider treats all these ideas and emphasizes the BLM's steps to conserve additional vegetation types via wilderness areas in a landscape where national parks are centered chiefly on scenic geological features.

Cooperrider shares many of Noss's concerns with size, fragmentation, and the need for a landscape-scale approach to conservation. Indeed, his attention to the importance of size, the careful management of seminatural buffer areas, and the riparian strips of crucial biological significance in the West make him an exponent of what may become known as "New Rangeland Management" (somewhat like Jerry Franklin's (1989) "New Forestry"). He also echoes Hal Salwasser's call for further interdisciplinary and interagency cooperation, stressing the need for more federal leadership in conservation. Unlike Salwasser, however, Cooperrider explicitly acknowledges the tensions now evident between conventional approaches to land and wildlife management and the emerging discipline of conservation biology.

In Chapter 4, Hal Salwasser summarizes how the USDA Forest Service is beginning to approach the issue of diversity on the 191-million-acre National Forest System. His concerns include the scope of such activities (reassuringly broad) as well as the need for more scientific knowledge and its effective dissemination. Throughout, Salwasser wisely stresses the need for integrated approaches that encompass a variety of spatial scales and involve all neighboring land managers.

The vast domain of the national forests ensures that they contain a great many sensitive species and communities. Moreover, the 1976 National Forest Management Act stipulated that all national forests develop long-term management plans, consider environmental impacts, and devise schemes that will "preserve and enhance" the biological diversity of each forest. This mandate has placed them squarely in the center of many controversies regarding biodiversity. The Forest Service's traditional preoccupation with commodity production and fulfilling congressional mandates to provide multiple uses has collided noisily with awakened public concerns regarding habitat loss, threats to diversity, and timber policies that often promote logging even in areas where it is environmentally damaging or uneconomic.

How is the Forest Service mobilizing to deal with the acute biodiversity issues it faces? Such issues are increasingly public: population declines of red-cockaded woodpeckers in the Southeast, the need to provide for large and connected old-growth habitats for spotted owls in the Northwest, concerns over the effects of increased edge and habitat fragmentation on declining forest-interior songbirds, and increasing ungulate populations in the East and Midwest. Like many agencies, the Forest Service is in transition with regard to implementing biodiversity conservation. While it is true that management plans have been successfully coordinated among agencies, the Forest Service also testified before Congress against the proposed National Biological Diversity Conservation and Environmental Research Act with its practical remedies to promote coordination. If the "New Perspectives" program, with its intent to promote environmentally sensitive forestry, is to succeed, the Forest Service must begin to demonstrate greater commitment within and consistency among its programs.

Beyond the federal agencies, The Nature Conservancy has proved itself to be an agile and powerful private force for conservation. In the final chapter in this part of the book, Bennett Brown reviews this organization's three-pronged approach to conserving biological diversity: identification/design, protection, and management. In the absence of effective federal leadership, the Conservancy has developed and implemented its own innovative Natural Heritage program to assess and document various elements of diversity state by state. These databases were designed to aid the Conservancy's efforts to acquire and protect the areas of greatest biological importance. The Conservancy has demonstrated that it plays a unique role in conservation by nimbly using its power as a private organization to quickly acquire threatened lands and secure long-term protection through a variety of legal tools.

As our appreciation of the importance of scale, linkages, and natural disturbance has developed, the Conservancy has kept pace by placing greater emphasis on management and the need to protect larger and more connected natural areas. Brown relates this history and the Conservancy's recent interest in acquiring and managing extensive "bioreserves." Throughout its history, the Conservancy's leadership and emphasis on systematic and ecologically informed approaches have also set crucial examples. Thus it is hardly a coincidence that the BLM's pursuit of gap analysis was predicated on the existence and availability of the Conservancy's Natural Heritage databases. If, indeed, we now rest on the threshold of an era of informed and effective cooperation that dovetails private, state, and federal conservation efforts, it is due in no small part to the work of The Nature Conservancy.

AT THE TURNING POINT

We face a historic turning point in the development of conservation biology as a discipline. Tactical efforts to conserve single species and habitats *in situ* have met with limited success, leading us to ponder the importance of large reserved areas, metapopulation dynamics, and the role of natural disturbance regimes. At the same time, we witness the way in which social, economic, and political pressures continue to fragment and denature habitats at an accelerating rate. Clearly, the time has come for conservation biologists to join the authors writing here to press for integrated conservation strategies. The best prescriptions can only effect a cure if they are reliably and consistently implemented.

Consensus is now emerging within the conservation community regarding the steps needed to preserve biological diversity and the importance of integrating efforts to achieve these goals. Fortunately, conservation biologists share certain values regarding the importance of preserving diversity through the protection of large and connected areas. We also collectively recognize the need to carefully assess and manage diversity across local and regional scales. But despite this unanimity, our conservation efforts remain scattered and our victories limited. If we are to do more than fight rearguard actions against the growing assault of lost habitats, extirpated populations, and threatened species, much more forceful and coordinated activity will be necessary.

Public interest, our scientific tools, and the training of skilled personnel have progressed quickly, yet we still lack the appropriate political context in which to implement the unified conservation strategies outlined here. To surmount the many institutional and economic pressures favoring resource extraction over conservation, we desperately need not only federal funding but federal leadership. Only mandates like those provided by the proposed National Biological Diversity Conservation and Environmental Research Act (Blockstein 1988) and a strongly funded and independent National Institutes for the Environment can ensure that our lofty goals stand a chance of effective implementation.

REFERENCES

Blockstein, D. E. 1988. Biological diversity bill introduced. *Bioscience* 38:455.
Chase, A. 1986. *Playing God in Yellowstone*. New York: Atlantic Monthly Press.
Franklin, J. F. 1989. Toward a new forestry. *Am. Forests* (Nov./Dec.):1–8.
Janzen, D. H. 1983. No park is an island: increase in interference from outside as park size decreases. *Oikos* 41:402–410.

MacArthur, R. H., and E. O. Wilson. 1967. *The theory of island biogeography.* Princeton: Princeton University Press.

Meine, C. 1988. *Aldo Leopold, his life and work.* Madison: University of Wisconsin Press.

Newmark, W. D. 1987. A land-bridge island perspective on mammalian extinctions in western North American parks. *Nature* 325:430–432.

Petraitis, P. S., R. E. Latham, and R. A. Niesenbaum. 1989. The maintenance of species diversity by disturbance. *Quart. Rev. Biol.* 64:393–418.

Pickett, S.T.A., and J. Thompson. 1978. Patch dynamics and the design of nature reserves. *Biol. Conserv.* 13:27–37.

Solheim, S. L., W. S. Alverson, and D. M. Waller. 1987. Maintaining biotic diversity in national forests: the necessity for large blocks of mature forest. *Endangered Species Tech. Bull. Reprint* 4(8):1–3.

Soulé, M. E., and D. S. Simberloff. 1986. What do genetics and ecology tell us about the design of nature reserves? *Biol. Conserv.* 35:19–40.

Steen, H. K. 1976. *The U.S. Forest Service: a history.* Seattle: University of Washington Press.

1 Gap Analysis: Assessing Protection Needs

*J. Michael Scott, Blair Csuti,
and Steven Caicco*

THE THREAT OF mass extinctions as a consequence of global warming (Peters 1988) joins a growing list of assaults on planetary biodiversity. Other effects of overpopulation (habitat loss, pollution, overexploitation) have driven thousands of species toward the brink of extinction. The U.S. Fish and Wildlife Service currently lists over 1,000 endangered and threatened species, with several thousand more "candidates" on the waiting list. But this is just the tip of the iceberg. Recent estimates (Erwin 1988; Wilson 1988) indicate there are more than 30 million species on earth, but a quarter of them may not see the year 2010 (Norton 1988). Most are insects, many not yet described, that play critical roles in the function of natural ecosystems upon whose biogeochemical cycles we depend for indispensable ecosystem support services (Wilson 1987). The rescue of the current list of endangered species has proved difficult, risky, and expensive. What does this record foreshadow as tens of thousands more species become endangered over the coming years?

Clearly, it is inefficient to save selected species while allowing the natural communities and ecosystems that support them (along with myriad inconspicuous species) to deteriorate. It would be wiser, surely, to identify and manage functioning representatives of each ecosystem type for the maintenance of national biodiversity. While very localized species, likely to be missed by a network of biodiversity

15

management areas (BMAs), would still require individual protection programs, such an integrated conservation strategy would ensure that the vast majority of species never become endangered. Further, if we were certain that the nation's biological resources were secure, non-critical lands could be put to sustainable uses without imperiling native species and ecosystems. The challenge, then, is to build a geographic database of biodiversity to assess the effectiveness of present and proposed BMAs. This process of identifying unprotected species or communities has been described as "gap analysis" (Burley 1988; Scott et al. 1988). Our objectives here are to present the rationale for this new approach for stemming the tidal wave of extinctions threatening the world's natural heritage and to describe the methods used in the model programs in Idaho and Oregon.

DESCRIBING THE DISTRIBUTION OF BIODIVERSITY

The distribution of the resources we are addressing (plants, animals, vegetation cover) is often incompletely documented—a shortcoming that frustrates attempts to develop a national conservation strategy and has led to calls for a national biological survey (Kosztarab 1984; Wilson 1985). But the identification and classification of the nation's insect fauna alone would take years, during which time conservation opportunities would be foreclosed. Nor is there time or funding for the intensive field surveys necessary to document species distributions on the basis of verified specimen records.

The intensive collecting carried out by the Bureau of Biological Survey (predecessor of the U.S. Fish and Wildlife Service) in the late nineteenth and early twentieth centuries forms much of the basis for our current understanding of plant and animal distribution. Despite decades of fieldwork, the ranges of most species are still predicted from the locations of a few specimens and the distribution of habitat types for that species. For example, Bailey (1936:368) suggests that Preble's shrew (*Sorex prebeli*) "probably has a somewhat interrupted range in the mountains and high country of eastern Oregon" on the basis of the capture locations of three specimens.

Using the association between species and habitat types to predict distribution has a long history of successful application to many common species (Armstrong 1972; Hoffmeister 1986). As Baker

(1956:122) succinctly put it: "Mammals generally are confined to specific kinds of plant associations from which they derive either food or shelter or both. Once the investigator has learned the ecological preferences of a given kind of mammal, he can map the occurrence of that mammal by noting the occurrence of the plants." While habitat specificity varies between species and classes of vertebrates, habitat is a powerful predictor of the distribution of many smaller mammals and birds, as well as reptiles and some amphibians.

Direct large-scale documentation of the distribution of an entire fauna is rarely attempted. A recent survey of Hawaiian forest birds (Scott et al. 1986) spanned seven years and recorded a quarter million observations at 10,000 field locations. Application of these methods to the continental United States is clearly impractical. While very rare or localized species are best dealt with on a locality-by-locality basis, we believe that, within known range boundaries, the fine-scale distribution of most species can be predicted from a knowledge of habitat preferences and the distribution of vegetation types.

While an intensive and thorough biological survey is a worthy undertaking, there is a growing consensus that a timely conservation strategy will have to be based on indirect indicators of biodiversity (Roberts 1988). Vegetation, vertebrates, and butterflies are the groups whose distribution is best documented (Pyle 1982; Scott et al. 1987). An analysis of their distribution relative to current nature reserves, national parks, wilderness areas, and the like would yield an assessment of the current level of protection of national biodiversity (Scott et al. 1987, 1988, 1989) and suggest management alternatives to fill the gaps in the network of reserves and BMAs.

VEGETATION AS WILDLIFE HABITAT

Vegetation is the most widely used indirect indicator of biodiversity (Crumpacker et al. 1988). Since it forms the basis for predicting animal distributions, an accurate vegetation map is the first priority of a gap analysis. Kuchler (1964; 1988) has led the way in mapping the distribution of "potential natural vegetation" (PNV) in the United States. The vegetation types depicted on Kuchler's 1964 map of "potential natural vegetation of the coterminous United States" (scale 1:3,168,000) "provide the only assessment of major, aboveground, terrestrial, and wetland ecosystem diversity that describes the entire United States in reasonable detail" (Crumpacker et al. 1988). In these places PNV is a valuable surrogate for biodiversity. Significant areas, however, have been converted to other cover types (such as cropland and pasture, urban and industrial areas, and early successional stages of forests). Wildlife responds to "real" or current vegetation rather than PNV. Thus it is possible to use county-of-occurrence information and a current vegetation map to predict the range of a species. Because vegetation forms the basis for predicting range maps, the first step in gap analysis is to create a vegetation map. In the pilot project in Idaho, we used a minimal mapping unit of 640 acres and a scale of 1:500,000.

METHODS FOR MAPPING VEGETATION

Because our time and money were limited, we chose to compile the map of Idaho vegetation primarily from existing sources. About two-thirds of Idaho is public land under the administration of the federal government. Most of this land is managed either by the USDA Forest Service (USFS) or the Bureau of Land Management (BLM). For most of these areas, maps of local vegetation were available from these agencies.

The most recent maps were those produced by the BLM during the past decade as part of the environmental assessment of their management actions. Maps from planning documents were reduced to our working scale of 1:500,000 and then traced onto mylar. In order to maintain a consistent level of discrimination throughout the areas mapped, it was often necessary to generalize the information from source maps of greater detail.

Because the Forest Service's current management practices emphasize site potential, comprehensive information on the actual vegetative cover of lands under its jurisdiction was not available. For these areas, we used timber type maps at a scale of 1:31,680 from the period 1950–1970. These maps were produced through interpretation of aerial photographs. In most cases, types were named on the basis of a single dominant tree species (more than 50 percent canopy coverage); in a few cases, codominant tree species were indicated.

The large scale of these maps, compared to our working scale, made it necessary to compress a great deal of information. This was accomplished visually. Contiguous maps were laid out by township, timber types were color-coded where necessary, and the major forest types within the township were delineated. This delineation was then transferred to mylar at our working scale. There are numerous areas of Idaho for which there are no local vegetation maps. Most of these areas. are in agricultural uses and were mapped as cropland or pastureland. The boundaries of these areas were delineated using a variety of sources, including soil surveys and topographic and geological maps. In nonagricultural areas adjacent to federal lands for which vegetation maps were available, type boundaries were extrapolated. A final edit of the map for agricultural boundaries and areas of recent timber harvest was performed using false-color infrared paper prints of Landsat MSS imagery.

As a result of our work in Idaho, we have modified the mapping procedures for vegetation. Habitat mapping can be done using digital image classification or manual interpretation of satellite imagery.

Both habitat mapping methods have advantages and disadvantages. With manual mapping, the large polygons drawn by interpreters may conform better to habitat classification systems. Moreover, they cost less to store and analyze than digital mapping and, in addition, photo-interpretation requires less training and technology. On the other hand, digital classification can depict greater spatial detail while avoiding errors in boundary placement and digitizing. Digital image classification also retains information on the spatial diversity occurring in habitat mosaics, whereas this diversity is usually generalized to one or another habitat type by photointerpreters.

CHOOSING A MAPPING STRATEGY

The choice of mapping strategy may well be dictated by time and budgetary constraints and is also somewhat region-specific. This, plus the fact that some states may want to produce habitat maps for uses other than gap analysis, makes it impossible to set uniform mapping procedures. In our experience, manual interpretation of specially pre-processed thematic mapping (TM) prints provides an accurate product in vector format that is easily merged with other variables such as landownership and land management. (These linear features are not well suited to the raster data structure of digital satellite data.) Much detail is lost using this approach, but the resulting map polygons are likely to be coherent landscapes well suited to regional analysis and planning. Furthermore, the manual maps are useful in segmenting digital satellite data for more detailed local analyses using digital classification.

The use of different mapping strategies by adjacent states raises the problem of edge-matching maps for broad regional analyses. In our experience, even edge-matching maps produced by the same method and classification system, but using different satellite scenes, can be difficult. The problem of edge-matching maps produced by digital and manual methods is clearly formidable.

PREDICTING ANIMAL DISTRIBUTIONS

The boundaries of traditional range maps enclose known records of occurrence. Unexplored regions may be omitted in error and, due to the small scale of most range maps, areas of inappropriate habitat types may be included. Many conservation and land-use planning decisions require more detailed information on species distributions. Because our 1:500,000 scale map of Idaho depicts polygons as small as 259 hectares (640 acres)—the average polygon size is 11,285 hectares for 118 actual vegetation types—a much finer resolution of preferred habitat types is possible. We used existing descriptions of habitat relationships to predict the presence or absence of native terrestrial vertebrates (with the exception of a few microhabitat specialists) in each vegetation cover type. Since the landscape-scale vegetation polygons include an unpredictable dispersion of seral stages and microhabitat features, we predict a species present within a polygon but not at particular locations within the polygon. And since the timing, degree of dispersal, and habitat preferences of wintering birds are difficult to predict, we considered only breeding bird ranges. Because of the complexity and scale of the vegetation map and the large number of vertebrate species, we are using an ARC/INFO Geographic Information System (GIS) to record and analyze these geo-based "themes."

Current small-scale Idaho distribution maps are available only for reptiles and amphibians (Nussbaum et al. 1983). These maps define the area within which a species is likely to occur in proper habitat. Very small scale distribution maps are available for mammals (Hall 1981), but since they depict continental distribution there are few records from Idaho. The distribution of Idaho birds by 1 × 1 degree blocks of latitude and longitude ("latilongs") is also available. A final guide to establishing range limits is a statewide county-of-occurrence database for all vertebrate species developed by the Idaho Natural Heritage program.

We use the GIS to superimpose two data sets for each species: distribution by county and association with vegetation cover type. The intersection of these data sets becomes the predicted distribution for a species. Initial comparisons with maps drawn by conventional methods are favorable. Each map is reviewed by local experts and further validated by comparison with museum specimen localities. Field validation of the distribution maps is planned for Idaho in 1992.

GAP ANALYSIS

The set of distribution maps, the vegetation map, a map of biodiversity management areas, and a map of classes of landownership form the data set for the gap analysis. Since the ranges of rare plants and animals cannot be accurately predicted by the distribution of vegetation cover types, specific localities at which these are known to occur are available from the Idaho Natural Heritage program and are included in the analysis as a separate data layer. Biodiversity management areas are defined as areas capable of being managed for the maintenance of native species and vegetation. Since management activities vary for different designations and ownerships, we qualify BMAs with a scale developed by The Nature Conservancy (TNC):

- LEVEL 1: total protection of native communities (national parks, TNC preserves)
- LEVEL 2: partial protection of native communities (wilderness areas, BLM areas of critical environmental concern)
- LEVEL 3: no protection (most private land, nondesignated public lands)

Although a number of different queries are possible regarding the management status of different animal species and vegetation types, the first result of developing a GIS database of vegetation is the creation of baseline information about the distribution and quantity of natural vegetation and its associated species. Since Landsat imagery is used to determine the current extent of cultural land-cover types (agricultural land, urban areas, forest fragmentation patterns), future updates will allow quantification of land use trends on a statewide or regional scale. Landscape patterns—such as the location and width of corridors of natural vegetation between BMAs—can also be mapped and monitored. We believe that, at a minimum, land use status should be updated at ten-year intervals (and more frequently in rapidly changing landscapes).

The first stage of a gap analysis is elementary: Which species and vegetation types are represented on BMAs? An initial gap analysis in south-central Idaho revealed that only nine of twenty-three vegetation types occurred on BMAs, for example, but none had more than 2.5 percent of their extant area in protected status. Further, BMAs covered only 0.6 percent of the entire study area, even though 41 percent of the land was in federal ownership and the largest BMA was smaller than 2,000 hectares.

One of the most widely accepted concepts of conservation biology is that of a minimum viable population (MVP; see Shaffer 1981). While the exact number of individuals required for long-term population persistence must be determined by a consideration of a number of risk factors, called a population viability analysis (Gilpin 1989), the relationship between density and area is a guide to minimum area requirements of target species. This minimum size can be compared with existing BMAs to assess their contribution to long-term management goals. A similar concept—minimum dynamic area (Pickett and Thompson 1978)—defines the area required for the expression of normal disturbance regimes necessary to maintain a landscape in a mosaic of characteristic successional stages. This concept offers another assessment of the adequacy of the BMA network.

It is unlikely that all species will be adequately represented in the BMA network. Certain habitats, especially those near growing urban centers or areas with resource exploitation potential (farming, timber and mineral extraction, type conversion), are particularly vulnerable to habitat degradation and fragmentation. This near certainty suggests that without preemptive management action, inadequately protected species may eventually reach endangered status. With the benefit of this foresight, a gap analysis can identify the fewest number of potential BMAs that would efficiently capture the greatest number of unrepresented species and vegetation types. In retrospect, the current plight of plants and animals native to California's southern San Joaquin Valley could have been avoided if past generations had set aside a viable representative of this currently endangered ecosystem. Several species of plants and animals are unique to this area, including the San Joaquin kit fox (*Vulpes macrotis mutica*), the giant kangaroo rat (*Dipodomys ingens*), and the blunt-nosed leopard lizard (*Gambelia silus*). Half a century ago these species were common, but once the California Aqueduct brought inexpensive irrigation water to the valley, most of their habitat was converted to agriculture. Because of this habitat loss, all three animals are now federally listed endangered species. It would have been better for the animals and cheaper for humans if we had avoided this crisis through early land use planning. Gap analysis provides the information needed to carry out such planning in the future.

While gap analyses are conducted on a state-by-state basis, a more biologically defensible assessment of the protection afforded biodiversity can be conducted within bioregions. Each biotic area (Clark 1937; Davis 1939) tends to have a natural assemblage of species. Five such areas have been identified in Idaho. They largely correspond to the more modern "ecoregion" concept (Omernik and Gallant 1986). By searching for areas of high habitat heterogeneity, the GIS can identify areas within each bioregion especially rich in species and vegetation cover types. Areas rich in local endemics are particularly efficient conservation targets (Terborgh and Winter 1983; Scott et al. 1987). Many federal land management units in the western United States are tens of thousands of hectares in size. Working at this scale, it is possible to identify species-rich management units as potential biodiversity management areas. Not surprisingly, such areas are often characterized by high topographic diversity.

APPLICATIONS AT LOCAL AND REGIONAL SCALES

The scale of our initial gap analysis was selected to capture landscape features: mosaics of ecosystems and the linkages between them. Once areas of special interest are identified, field surveys and the designation of management boundaries are carried out at a larger scale (1:24,000). There is a direct application of this step-down process to land use planning at the county level. Future changes in urban growth boundaries, designated agricultural zones, greenbelts, and transportation corridors, to name but a few, are best visualized at a landscape scale but must be applied to individual parcels of land. It is especially clear in densely populated areas like southern California that unplanned development of wildlands can lead to costly and contentious conflicts between environmentalists and developers. Once they reach a crisis stage, these conflicts are usually resolved in a confrontational climate, leading to much expense and dissatisfaction for all parties. Less developed regions of the country have the option of using GIS databases of natural resources to plan development patterns through relatively inexpensive zoning regulations or other land use planning mechanisms.

Plants, animals, and vegetation pay no heed to political boundaries (except for those situated on major biogeographic barriers to dispersal). While a statewide gap analysis is useful, a regional analysis is more valuable. Since major biotic areas often transcend state boundaries, it would be redundant for each state to attempt protection of a biological region shared by many states. An analysis of patterns of species richness by bioregion is likely to identify several high-priority sites that can then be ranked from a regional rather than a state perspective.

The western United States is a particularly promising area for gap analysis. Over half the Kuchler PNV types occur west of the 100th meridian. Perhaps more important, the topographic relief and climatic diversity of the West provide a natural laboratory for evolution. The state of California alone is home to sixty-one endemic species of vertebrates. The Rocky Mountains represent not only the beginning of a western flora and fauna but also a change in landownership from predominantly private to largely public lands, most of them undeveloped. The opportunity to maintain national diversity through changes in management prescriptions on public lands exists primarily in the West. Through cooperation with private landowners and conservation groups, a set of biodiversity management areas, coupled with selected endangered species reserves, could stem future extinctions in the region and serve as a model for international planning. We believe that Geographic Information System gap analysis of the distribution and status of natural resources is the appropriate technology to meet the challenge of locating these management areas. This goal could be achieved in six years at a cost of less than a penny an acre.

REFERENCES

Armstrong, D. M. 1972. Distribution of mammals in Colorado. Mus. Nat. Hist., Univ. Kansas, Monogr. no. 3:1–415.

Bailey, V. 1936. The mammals and life zones of Oregon. *N. Am. Fauna* 55:1–416.

Baker, R. H. 1956. Mammals of Coahuila, Mexico. Univ. Kansas Publ., Mus. Nat. Hist., 9(7):125–335.

Burley, F. W. 1988. Monitoring biological diversity for setting conservation priorities. Pp. 277–230 in *Biodiversity* (E. O. Wilson, ed.). Washington: National Academy Press.

Clark, H. W. 1937. Association types in the north coast ranges of California. *Ecology* 18:214–230.

Crumpacker, D. W., S. W. Hodge, D. Friedley, and W. P. Gregg, Jr. 1988. A preliminary assessment of the status of major terrestrial and wetland ecosystems on federal and Indian lands in the United States. *Conserv. Biol.* 2:103–115.

Davis, W. B. 1939. *The Recent mammals of Idaho*. Caldwell, Idaho: Caxton Printers.

Erwin, T. L. 1988. The tropical forest canopy: the heart of biotic diversity. Pp. 123–129 in *Biodiversity* (E. O. Wilson, ed.). Washington: National Academy Press.

Gilpin, M. 1989. Population viability analysis. *Endangered Species Update* 5(10):15–18.

Hall, E. R. 1981. *The mammals of North America*. 2nd ed. New York: Wiley.

Hoffmeister, D. F. 1986. *Mammals of Arizona*. Tucson: University of Arizona Press.

Kosztarab, M. 1984. A biological survey of the United States. *Science* 233:443.

Kuchler, A. W. 1964. Potential natural vegetation of the coterminous United States: manual to accompany the map. Am. Geog. Soc., Spec. Publ. no. 36.

Kuchler, A. W., and I. S. Zonneveld. 1988. *Vegetation mapping*. Dordrecht: Kluwer.

Norton, B. 1988. Commodity, amenity and morality: the limits of quantification in valuing biodiversity. Pp. 200–205 in *Biodiversity* (E. O. Wilson, ed.). Washington: National Academy Press.

Nussbaum, R. A., E. D. Brodie, and R. M. Storm. 1983. *Amphibians and reptiles of the Pacific Northwest*. Moscow: University Press of Idaho.

Omernik, J. M., and A. L. Gallant. 1986. Ecoregions of the Pacific Northwest. U. S. Environmental Protection Agency, Environmental Research Laboratory, Corvallis, Oregon.

Peters, R. L. II. 1988. The effect of global climate change on natural communities. Pp. 450–461 in *Biodiversity* (E. O. Wilson, ed.). Washington: National Academy Press.

Pickett, S.T.A., and J. N. Thompson. 1978. Patch dynamics and the design of nature reserves. *Biol. Conserv.* 13:27–37.

Pyle, R. M. 1982. Butterfly ecogeography and biological conservation in Washington. *Atala* 8:1–26.

Roberts, L. 1988. Hard choices ahead on biodiversity. *Science* 241:1759–1761.

Scott, J. M., B. Csuti, J. E. Estes, and H. Anderson. 1989. Status assessment of biodiversity protection. *Conserv. Biol.* 3:85–87.

Scott, J. M., B. Csuti, J. D. Jacobi, and J. E. Estes. 1987. Species richness: a geographic approach to protecting future biological diversity. *Bioscience* 37:782–788.

Scott, J. M., B. Csuti, K. Smith, J. E. Estes, and S. Caicco. 1988. Beyond endangered species: an integrated conservation strategy for the preservation of biological diversity. *Endangered Species Update* 5(10):43–48.

Scott, J. M., S. Mountainspring, F. L. Ramsey, and C. B. Kepler. 1986. Forest bird communities of the Hawaiian Islands: their dynamics, ecology, and conservation. *Studies in Avian Biology*, no. 9, Cooper Ornithological Society.

Shaffer, M. 1981. Minimum population sizes for species conservation. *Bioscience* 31:131–134.

Terborgh, J., and B. Winter. 1983. A method for siting parks and reserves with special reference to Colombia and Ecuador. *Biol. Conserv.* 27:45–58.

Wilson, E. O. 1985. The biological diversity crisis. *Bioscience* 35:700–706.

————. 1987. The little things that run the world (the importance and conservation of invertebrates). *Conserv. Biol.* 1(4):344–346.

————. 1988. The current state of biological diversity. Pp. 3–18 in *Biodiversity* (E. O. Wilson, ed.). Washington: National Academy Press.

2 Landscape Connectivity: Different Functions at Different Scales

Reed F. Noss

HABITAT FRAGMENTATION is considered by many biologists to be the single greatest threat to biological diversity. One strategy offered to counter the fragmentation problem is that of landscape linkages, usually understood as linear corridors of habitat that physically connect larger habitat patches in a landscape mosaic. Here I address the broad concept of "connectivity," which involves linkages of habitats, species, communities, and ecological processes at multiple spatial and temporal scales. I argue that because ecological processes and elements of biological diversity occur at a variety of scales, a comprehensive strategy to conserve these processes and elements must also encompass a diversity of scales.

Many of the most significant human effects on biodiversity involve changes in the connectivity of biological phenomena. Human activities may either reduce or increase connectivity. We have created artificial barriers to species dispersal in many instances, whereas in other cases we have eliminated natural barriers. In the former situation, isolated populations become more vulnerable to extinction due to reduced access to resources, genetic deterioration, increased susceptibility to environmental catastrophes and demographic accidents, and other problems (Harris 1984; Soulé 1987). In the latter situation, we make it easier for exotic organisms to invade indigenous communities. The end result of this process is a homogenization of floras and faunas, an example of which Crosby (1986) has called "ecological imperialism": the conversion to a biological neo-Europe of regions settled by people of European descent.

PROCESSES OF MOVEMENT AND DISPERSAL

In biology, motion occurs at many levels of organization. Of most relevance to conservation biology is the movement of organisms across landscapes (and seascapes) and the movement of alleles (genes) within and among populations of organisms. Many animals make daily and seasonal movements to meet life-history needs, and they depend upon corridors or "stepping-stones" of appropriate habitat to do so. Dispersal, on the other hand, refers to the movement of organisms away from their place of origin; the movement of alleles due to dispersal of gametes or offspring is called gene flow (Brown and Gibson 1983). It is particularly critical that avenues for dispersal be maintained in times of climate change, when organisms must track shifting habitat conditions if they are to survive.

A striking feature of life on earth is its lack of uniformity in distribution (Cox and Moore 1985). Biodiversity is a dynamic consequence of dispersal and isolation. If we look at any local site on earth, except for oceanic islands and a few islandlike continental habitats, the overwhelming majority of species we see came from somewhere else. Only a small number of species, if any, evolved there *in situ*. As we enlarge our area of observation, a greater number of species are endemic to that area—that is, they are found nowhere else. The island continent of Australia has a large proportion of endemic species (and now many human-introduced exotics). In many cases, endemics are thought to have evolved within the areas in which they are found today. In other (probably fewer) cases, an endemic species may once have been widespread but then contracted in distribution to its present range. For any area, including islands, endemics evolved from ancestors that originated elsewhere, ultimately going back to the primordial and unknown place where life began. Thus dispersal—followed by at least partial isolation and subsequent differentiation into new species or genetically distinct populations—generates biodiversity.

Of great import to conservation biology is the fact that organisms differ in their dispersal capacities. Most birds, bats, and flying insects are mobile, whereas terrestrial snails and flightless insects are quite sedentary. Plants that have their seeds dispersed by birds or wind may be very mobile, whereas forest wildflowers dispersed by ants are not. Furthermore, a corridor to one species may be a barrier to another. A strip of forest through agricultural land, for example, would provide a corridor for dispersal of forest animals that avoid clearings, but it would form a barrier to such species as the meadow vole. A river facilitates movement of fish but may form a barrier for many terrestrial organisms. Related species or subspecies often are found on opposite sides of major rivers (Hesse et al. 1937). Some tropical birds, despite their powers of flight, have psychological barriers to crossing even narrow water gaps (Diamond 1975).

Different species disperse through different types of habitat. Biogeographers usually limit their use of the term "corridor" to broad, internally heterogeneous swaths of habitat that permit the direct spread of many or most taxa from one region to another (Brown and Gibson 1983). "Filters" (Simpson 1940), on the other hand, are dispersal routes containing a more limited habitat spectrum through which certain species pass and others are excluded. A clearcut, for example, is a filter that some forest mammals (such as black bear) may cross but others such as pine marten avoid (Soutiere 1979). When organisms are faced with crossing totally unsuitable habitats— terrestrial amphibians making their way to oceanic islands, for example, or a Florida panther trying to cross downtown Miami—such chance movement is called "sweepstakes dispersal." Our knowledge of what constitutes a barrier or a corridor for various species is extremely limited, however. We do not even have adequate models for dispersal of animals or plants through heterogeneous environments; most models of dispersal (for example, based on diffusion equations) assume a homogeneous environment and a random walk (Lubina and Levin 1988).

The conservation strategy that emerges from these considerations is one that seeks to optimize the width and variety of natural habitats in landscape linkages so that the full spectrum of native species will be able to move between natural areas in the regional landscape (Noss and Harris 1986; Noss 1987a). Narrow corridors, or corridors comprising only one kind of habitat, are less useful. A narrow corridor may be entirely edge habitat that is not used by forest-interior species or subjects them to high rates of predation (Wilcove et al. 1986). Although riparian forests are often suggested as dispersal corridors, they may not be suitable for species of dry upland habitats (Forman 1983). An ideal corridor might encompass the entire topographic gradient and habitat spectrum from river to ridgetop. But in some landscapes, wide corridors are no longer available and cannot, in the short term, be restored. In such cases, multiple corridors ("networks") may collectively encompass the full range of native habitats and also ensure against destruction of single corridors by catastrophic disturbances (Forman and Godron 1981).

The flip side of the corridor strategy is to minimize barriers. Urban areas, agricultural fields, clearcuts, and roads make dispersal difficult or dangerous for many forest-interior and wide-ranging species (Harris and Gallagher 1989). Roads are particularly troublesome, as they now cover much of the landscape and are filters through which many organisms cannot pass. Oxley et al. (1974) found that small forest mammals in southern Canada rarely ventured onto road surfaces when the width of the road clearance exceeded 20 meters. In Germany, several species of forest-dwelling carabid beetles and mice rarely or never crossed two-lane roads; even a small, unpaved forest road closed to public traffic was a barrier (Mader 1984). Surprisingly, even mammals of open habitats such as deserts may rarely cross highways (Garland and Bradley 1984). For species restricted from crossing roads, a road effectively fragments a population into subpopulations that are smaller and more vulnerable to extinction from any number of random or deterministic processes.

Climate change, whether natural or anthropogenic, intensifies the deleterious effects of roads and other artificial barriers. Many animals (and the plants they disperse) are simply unable to migrate along with their habitat conditions where dispersal barriers are insurmountable (Peters and Darling 1985). Even mobile species such as large mammals must run a gauntlet of roads, developments, and associated mortality risks to get from point A to point B. Closing roads, minimizing new road construction, and building underpasses and viaducts are some of the strategies suggested to alleviate this problem (Noss 1987b; Harris and Gallagher 1989).

SPATIOTEMPORAL SCALES OF CONCERN

An important realization of modern conservation biology is that conservation problems exist at many different spatiotemporal scales and levels of biological organization. Similarly, conservation strategies are applicable at multiple scales (Noss 1991). It is critical, of course, that the proposed solution match the scale and nature of the problem. Maintaining wooded fencerows may help white-footed mice travel among woodlots, but it will not do much for wolves. A strategy that maintains biodiversity in the short term (say a few years) may fail to preserve viable populations and ecological integrity over a longer time span.

What is the optimal scale for planning and managing habitat connectivity? It depends on the biota under consideration and the goals of the conservation strategy.

The Fencerow Scale. Gray Merriam and his students at Carleton University in Ontario have convincingly demonstrated the value of fencerows to small mammals in agricultural landscapes. Local extinctions of white-footed mice and chipmunks are common in small, isolated woodlots, but fencerows permit frequent recolonization and the metapopulation persists (Wegner and Merriam 1979; Fahrig and Merriam 1985; Henderson et al. 1985; Merriam 1988). Fencerows appear to allow mice to move between woodlots with less risk of predation than when they cross open fields. Even birds may depend on habitat corridors. In the Netherlands, the number of forest-interior birds in woodlots is significantly related to the density of connecting elements such as wooded banks and tree rows (Van Dorp and Opdam 1987). To avoid predation in fall from migrating hawks, blue jays in Wisconsin fly over fencerows instead of open fields and dive into cover if attacked (Johnson and Adkisson 1985). Many species of birds and mammals (as well as a snake) have been noted to use hedgerows for movement across a landscape, whereas the same hedgerows block movement of other animals such as cows (Forman and Baudry 1984). A disadvantage of fencerows and hedgerows is that they are narrow and therefore entirely edge habitat—thus forming "line" corridors instead of "strip" corridors (Forman and Godron 1981). Some forest-interior species are unable to use these narrow corridors, and weedy, opportunistic species of edge habitats may benefit at their expense (Simberloff and Cox 1987).

The Landscape Mosaic Scale. We usually think of corridors in terms of distinct linear strips from one habitat patch to another. But as noted above, the term "corridor" has a much broader meaning. Any area of habitat through which an animal or plant propagule has a high probability of moving may be considered a corridor or linkage. Wide-ranging mammals such as bears and mountain lions require such corridors to meet their daily needs for food, water, and shelter and may roam tens of miles in a 24-hour period. Riparian and topographic ridge systems are followed routinely by these species (Harris 1985). Amphibians and migratory ungulates use corridors for seasonal movements across the landscape mosaic, as do other animals that require a variety of habitats for different stages in their life cycles. Connectivity of the landscape mosaic is absolutely necessary for these species to survive. Furthermore, disturbances periodically make parts of the landscape uninhabitable. Corridors fulfill a "fire escape" function by permitting animals to flee disturbance; they then aid in the recolonization of the recovering site by plants and animals.

Another kind of connectivity has to do with continuity of habitats and processes along environmental gradients. Too often ecologists and conservationists have looked at habitats as separate entities, whereas in reality they are interacting, functional components of the landscape ecosystem (Noss 1987c). Since plant species usually are distributed independently along gradients, species diversity can be appreciated only by considering the gradient as a whole. Disturbances and other ecological processes interact with gradients of soil and moisture to determine the distribution of species. In pinelands of the Gulf coastal plain, for example, fires sweep down gradual slopes and prune back wetland shrubs that otherwise would encroach from adjacent swamps. In the process, these fires maintain an open herb-bog community with an extremely diverse flora, including many insectivorous plants. If fires are suppressed, or if firelanes disrupt the hydrology of the slope-moisture gradient, its unique flora is destroyed (Noss and Harris 1989). Thus connectivity of process is just as important as connectivity of habitat.

FIGURE 2.1 **A Proposed Network of Preserves, Buffer Zones, and Corridors for Florida.**

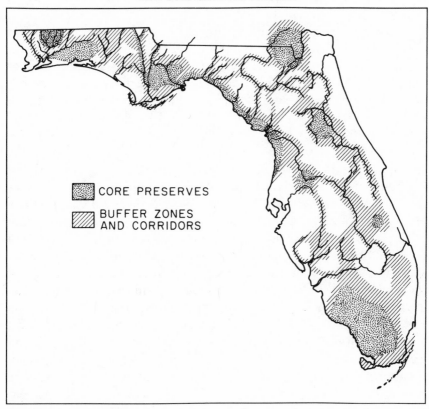

CORE PRESERVES

BUFFER ZONES AND CORRIDORS

Adapted from Noss (1987b).

Regional and Continental Scales. A remarkable development in conservation over the last decade has been the emergence, in many parts of the world, of ambitious strategies to connect nature reserves into regional networks (Harris and Gallagher 1989). An example of a comprehensive regional network is that proposed for the state of Florida plus Okefenokee Swamp in adjacent Georgia (Figure 2.1). Noss and Harris (1986), Harris (1988), and Harris and Gallagher (1989) have discussed many of the specific linkages in this system. Although the statewide network is admittedly an idealistic strategy, particularly in a state with over 1,000 new human residents each day, it has been endorsed by several environmental groups and, in principle, by the Florida Department of Natural Resources. The state of Florida, the federal government, and The Nature Conservancy have acquired several key linkages in the network, and other corridor acquisitions are pending (Noss 1987b).

There is a need to think even bigger than statewide networks. This need is especially evident when we enlarge our temporal view to centuries and millennia. Changes in climate in the near future may be much more rapid than most past changes, perhaps at the scale of decades. Even natural rates of climate change are a threat to species in fragmented landscapes. Although dispersal barriers exist in nature, habitat fragmentation has greatly increased the number of barriers for most native species. In the past, biogeographic corridors on massive scales, such as the Bering land bridge, the Isthmus of Panama, and the Gulf coastal savanna, have been critical determinants of present patterns of biodiversity (Simpson 1940, 1965; Brown and Gibson 1983; Harris 1985). Conservation biologists have proposed the establishment of corridors on regional and continental scales to permit the shifting of floras and faunas in response to global climate change (Hunter et al. 1988; Peters 1988). Suggested regional linkages include a network of protected areas paralleling the Appalachian Trail (Sayen 1987; Hunter et al. 1988) and a network of national parks, national forests, and national wildlife refuges in the western United States (Salwasser et al. 1987). Strong differences of opinion exist on how much human activity (such as clearcutting and roads) can exist within units of a network without seriously limiting its viability for wildlife. No longer is there any significant argument, however, over the need for such networks.

TO ENHANCE OR NOT TO ENHANCE

So far I have discussed the movement and dispersal of organisms as something that should be encouraged. The problems addressed in this volume relate to fragmentation of habitats and other artificial limitations of natural movement patterns. But at the same time human activity has restricted opportunities for movement of many taxa, it has abetted the movement of others. Floras and faunas that once were distinct and endemic are now dominated by exotics (nonnative species) and cosmopolitan weeds. The cause of this biological pollution has been twofold: human transportation that facilitates the spread of certain species far beyond their natural dispersal capacities and habitat modification that favors weedy invaders, such as annual plants, over native flora and fauna (Elton 1958; Mooney and Drake 1986). The same road that restricts movement of certain animal species may encourage movement of others. The meadow vole, for example, rapidly extended its range into central Illinois when a superhighway with grassy verges was constructed (Getz et al. 1978), and many weedy and exotic plants spread along roadsides. Monocultures, such as tree plantations, favor the spread of insect and fungal pests (including many exotics) far beyond the rates typical of heterogeneous, old-growth landscapes (Norse 1990).

A curious thing about biological invasions is that they often result in a community with more species than the original (Mooney 1988; Vitousek 1988). Should conservationists therefore applaud the spread of exotics? Not if we recognize that enhancement of species richness at a local scale can mean loss of species richness at a global scale, as sensitive endemics are lost and weeds prosper everywhere. Many of the Mediterranean-climate regions of the world, which once had distinct endemic floras, are now dominated by the same species of plants. Ecosystem-level diversity declines as distinct communities are lost; the inevitable result is global homogenization (Mooney 1988). The lesson for conservationists is that limiting the dispersal of opportunistic, invasive organisms (especially exotics) may be as important as enhancing the dispersal of sensitive native taxa—if the goal is to maintain native biodiversity and ecological integrity. Thus a suitable goal may be to optimize the connectivity of natural habitats in the landscape while at the same time minimizing the connectivity of artificial habitats such as clearcuts, agricultural fields, and roadsides.

A VIABLE OPTION

The controversy surrounding the corridor issue has involved concerns that the wrong species (that is, weeds) might benefit and that the costs of establishing corridors are so high that money would be better spent on habitat containing concentrations of endangered species (Simberloff and Cox 1987). Much of this concern has evaporated with the realization that small, isolated reserves will not maintain viable populations of many taxa in the long term, that climate change demands that organisms be able to move large distances over time, and that society has sufficient funds to buy both endangered species habitat and corridors—if it can find the will (Noss 1987a; Harris and Gallagher 1989).

Much of the discussion about corridors leaves the impression that we are constructing something new on the landscape. We talk of "establishing" corridors rather than "maintaining" corridors. In fact, an engineering approach to corridors is necessary chiefly when roads are in the way and underpasses, tunnels, or viaducts must be constructed so that animals can pass underneath them. Habitat restoration is also needed in many cases, sometimes on a massive scale. But the corridor strategy is fundamentally an attempt to maintain or restore natural landscape connectivity, not to build connections between naturally isolated habitats.

Critics point out that conservationists have not proved that any of these corridor strategies will work—or, more precisely, that the null hypothesis of *no* effect of corridors has not been proved false. But null hypotheses in ecology are not straightforward. In this instance, two alternative null hypotheses are possible: there is no effect of corridors; or there is no effect of fragmentation (that is, of eliminating natural corridors). Testing the second null hypothesis is arguably the more prudent course, as it errs on the side of preservation. We know that natural landscapes are fundamentally interconnected and that connectivity declines with human modification of the landscape (Godron and Forman 1983; Noss 1987a). Furthermore, the null hypothesis of no effect of fragmentation has been tested and falsified, repeatedly, in many parts of the world. Indeed, the effects of fragmentation can be devastating (Wilcove et al. 1986). In general, then, we can reject the null hypothesis of no effect of fragmentation and consider maintenance of habitat corridors a viable option for conservation. We have a lot to learn, however, about the management of specific corridors for biodiversity and human use combined.

REFERENCES

Brown, J. H., and A. C. Gibson. 1983. *Biogeography.* St. Louis: Mosby.
Cox, C. B., and P. D. Moore. 1985. *Biogeography: an ecological and evolutionary approach.* Oxford: Blackwell.
Crosby, A. W. 1986. *Ecological imperialism: the biological expansion of Europe, 900–1900.* New York: Columbia University Press.

Diamond, J. M. 1975. The island dilemma: lessons of modern biogeographic studies for the design of natural preserves. *Biol. Conserv.* 7:129–146.

Elton, C. S. 1958. *The ecology of invasions by animals and plants.* London: Methuen.

Fahrig, L., and G. Merriam. 1985. Habitat patch connectivity and population survival. *Ecology* 66:1762–1768.

Forman, R.T.T. 1983. Corridors in a landscape: their ecological structure and function. *Ekologiya* (CSSR) 2:375–387.

Forman, R.T.T., and J. Baudry. 1984. Hedgerows and hedgerow networks in landscape ecology. *Environ. Management* 8:495–510.

Forman, R.T.T., and M. Godron. 1981. Patches and structural components for a landscape ecology. *BioScience* 31:733–740.

Garland, T., and W. G. Bradley. 1984. Effects of a highway on Mojave Desert rodent populations. *Am. Midland Nat.* 111:47–56.

Getz, L. L., F. R. Cole, and D. L. Gates. 1978. Interstate roadsides as dispersal routes for *Microtus pennsylvanicus. J. Mammalogy* 59:208–212.

Godron, M., and R.T.T. Forman. 1983. Landscape modification and changing ecological characteristics. Pp. 12–28 in H. A. Mooney and M. Godron (eds.), *Disturbance and ecosystems.* Berlin: Springer-Verlag.

Harris, L. D. 1984. *The fragmented forest.* Chicago: University of Chicago Press.

————. 1985 *Conservation corridors: a highway system for wildlife.* ENFO Report 85–5. Winter Park: Florida Conservation Foundation.

————. 1988. Landscape linkages: the dispersal corridor approach to wildlife conservation. VCR film. Gainesville: Florida Films.

Harris, L. D., and P. B. Gallagher. 1989. New initiatives for wildlife conservation: the need for movement corridors. Pp. 11–34 in G. Mackintosh (ed.), *Preserving communities and corridors.* Washington: Defenders of Wildlife.

Henderson, M. T., G. Merriam, and J. Wegner. 1985. Patchy environments and species survival: chipmunks in an agricultural mosaic. *Biol. Conserv.* 31:95–105.

Hesse, R., W. C. Allee, and K. P. Schmidt. 1937. *Ecological animal geography.* New York: Wiley.

Hunter, M. L., G. L. Jacobson, and T. Webb. 1988. Paleoecology and the coarse-filter approach to maintaining biological diversity. *Conserv. Biol.* 2:375–385.

Johnson, W. C., and C. S. Adkisson. 1985. Dispersal of beech nuts by blue jays in fragmented landscapes. *Am. Midland Nat.* 113:319–324.

Lubina, J. A., and S. A. Levin. 1988. The spread of a reinvading species: range expansion in the California sea otter. *Am. Nat.* 131:526–543.

Mader, H. J. 1984. Animal habitat isolation by roads and agricultural fields. *Biol. Conserv.* 29:81–96.

Merriam, G. 1988. Landscape dynamics in farmland. *Trends in Ecology and Evolution* 3:16–20.

Mooney, H. A. 1988. Lessons from Mediterranean-climate regions. Pp. 157–165 in E. O. Wilson (ed.), *Biodiversity*. Washington: National Academy Press.

Mooney, H. A., and J. Drake (eds.). 1986. *The ecology of biological invasions of North America and Hawaii*. New York: Springer-Verlag.

Norse, E. A. 1990. *Ancient forests of the Pacific Northwest*. Washington: Island Press/Wilderness Society.

Noss, R. F. 1987a. Corridors in real landscapes: a reply to Simberloff and Cox. *Conserv. Biol.* 1:159–164.

————. 1987b. Protecting natural areas in fragmented landscapes. *Nat. Areas J.* 7:2–13.

————. 1987c. From plant communities to landscapes in conservation inventories: a look at The Nature Conservancy (USA). *Biol. Conserv.* 41:11–37.

————. 1991. Issues of scale in conservation biology. In P. L. Fiedler and S. K. Jain (eds.), *Conservation biology: the theory and practice of nature conservation, preservation, and management*. New York: Chapman & Hall.

Noss, R. F., and L. D. Harris. 1986. Nodes, networks, and MUMs: preserving diversity at all scales. *Environ. Management* 10:299–309.

————. 1989. Habitat connectivity and the conservation of biological diversity: Florida as a case history. In *Proceedings of the Society of American Foresters*. Bethesda, Md.: Society of American Foresters.

Oxley, D. J., M. B. Fenton, and G. R. Carmody. 1974. The effects of roads on populations of small mammals. *J. Appl. Ecol.* 11:51–59.

Peters, R. L. 1988. Effects of global warming on species and habitats: an overview. *Endangered Species Update* 5(7):1–8.

Peters, R. L., and J.D.S. Darling. 1985. The greenhouse effect and nature reserves. *BioScience* 35:707–717.

Salwasser, H., C. Schonewald-Cox, and R. Baker. 1987. The role of interagency cooperation in managing for viable populations. Pp. 159–173 in M. E. Soulé (ed.), *Viable populations for conservation*. Cambridge: Cambridge University Press.

Sayen, J. 1987. The Appalachian Mountains: vision and wilderness. *Earth First!* 7(5):26–30.

Simberloff, D., and J. Cox. 1987. Consequences and costs of conservation corridors. *Conserv. Biol.* 1:63–71.

Simpson, G. G. 1940. Mammals and landbridges. *J. Wash. Acad. Sci.* 30:137–163.

————. 1965. *The geography of evolution*. Philadelphia: Chilton.

Soulé, M. E. (ed.). 1987. *Viable populations for conservation*. Cambridge: Cambridge University Press.

Soutiere, E. C. 1979. Effects of timber harvesting on marten in Maine. *J. Wildl. Management* 43:850–860.

Van Dorp, D., and P.F.M. Opdam. 1987. Effects of patch size, isolation, and regional abundance on forest bird communities. *Landscape Ecol*. 1:59–73.

Vitousek, P. M. 1988. Diversity and biological invasions of oceanic islands. Pp. 181–189 in E. O. Wilson (ed.), *Biodiversity*. Washington: National Academy Press.

Wegner, J. F., and G. Merriam. 1979. Movements by birds and small mammals between a wood and adjoining farmland habitats. *J. Appl. Ecol*. 16:349–357.

Wilcove, D. S., C. H. McLellan, and A. P. Dobson. 1986. Habitat fragmentation in the temperate zone. Pp. 237–256 in M. E. Soulé (ed.), *Conservation biology: the science of scarcity and diversity*. Sunderland, Mass.: Sinauer.

Wilcox, B. A., and D. D. Murphy. 1985. Conservation strategy: the effects of fragmentation on extinction. *Am. Nat*. 125:879–887.

3 Conservation of Biodiversity on Western Rangelands

Allen Cooperrider

IN THE ELEVEN STATES west of the 100th meridian, rangelands comprise about 70 percent of the land. The postsettlement history of the West has had a common theme of exploitation and often destruction of biological resources, beginning with beaver trapping and followed a few decades later by widespread extirpation of bison. More recently, the decay of biodiversity in the West has involved primarily the loss of distinct populations and parts of ecosystems, rather than the extinction of entire species or ecosystems. But the accumulated losses of populations and ecosystem fragments could soon add up to a permanent disappearance of many species and communities (Ehrlich 1987). Conservation of biological diversity on these rangelands is desirable for ethical or aesthetic reasons. More compellingly, biological diversity provides the basis for a sustainable ecosystem, economy, and society. Thus it is essential to the health and welfare of the people of the West. This essay begins by reviewing current concepts of biological diversity and their evolution.

CURRENT CONCEPTS

Biological diversity has many connotations. The Office of Technology Assessment (OTA) has developed the following definition: "Biological diversity refers to the variety and variability among living organisms and the ecological complexes in which they occur. Diversity can be defined as the number of different items and their relative frequency. Thus, the term encompasses different ecosystems, species, genes, and their relative abundance" (U.S. Congress, Office of Technology Assessment 1987). Although numerous other definitions are available, most are characterized by explicit recognition of three levels of diversity (ecosystem, species, genes) and a broader description than the "species diversity" indices in vogue in the 1960s and 1970s that typically involved a quantitative diversity index.

As the OTA's definition of biodiversity has been criticized for not considering both form and function, a U.S. Bureau of Land Management (BLM) advisory group developed the following unofficial definition: Biodiversity is the aggregate of species assemblages (communities), individual species, and genetic variation within species and the processes by which these components interact within and among themselves; for purposes of classification, biodiversity can be divided into three levels: (1) community diversity (habitat, ecosystem), (2) species diversity, and (3) genetic diversity within species; all three levels change through time. This definition explicitly recognizes both form (structure and composition of ecological communities) as well as function (ecological processes such as succession; evolutionary processes such as speciation). Thus conserving biological diversity includes more than just recovery of endangered species or creation of preserves. It also encompasses maintaining ecological processes and preserving the capability of genes, organisms, and communities to evolve over time.

CURRENT INTEREST

Current concern over loss of biological diversity in this country has led to a major study by the OTA (U.S. Congress, Office of Technology Assessment 1987), numerous conferences and symposia resulting in books (Soulé and Wilcox 1980; Soulé 1986; Wilson and Peter 1988), and an emerging discipline known as conservation biology. The Society for Conservation Biology was founded in 1986 and publication of its journal, *Conservation Biology*, commenced in 1987. More recently, legislation to mandate conservation of biological diversity has been introduced in Congress (Blockstein 1988).

This rapid emergence of a new discipline has caused much unrest in some of the older conservation disciplines, particularly wildlife management (Capen 1989). Some have suggested that conservation biol-

FIGURE 3.1 Evolution of Concern for Species Groups
in the United States.

All Species	National Biological Diversity Conservation and Environmental Research Act
_____ 1990 _____	
Plants	Endangered Species Act (1973)
Fish	Fish Restoration and Management Act (1950)
Wild Birds	Migratory Bird Treaty Act (1918)
Wild Animals	Lacey Act (1900)
_____ 1900 _____	
Large (Huntable) Mammals, Birds, Fish	Early State Game Protection Laws (1700s)

ogy is merely an old profession (wildlife ecology and management) under a new name (Teer 1989), whereas others have fully embraced the new discipline (Thomas and Salwasser 1989).

Biological diversity is neither a new buzzword for an old concern nor an environmental fad; rather, concern over loss of biological diversity is a logical extension of a conservation movement that dates' back to at least the eighteenth century in this country (Cooperrider 1989). From this perspective, legislation such as the currently proposed "National Biological Diversity Conservation and Environmental Research Act" (HR 1268) represents a shift from a focus on certain species groups to a concern for all species (Figure 3.1). The proposed act also represents a similar expansion from a concern over particular ecosystems to a concern for all ecosystems (Figure 3.2). Finally, a concern for gene conservation is a logical result of our modern understanding of evolution and population genetics.

Conservation mandates as articulated in legislation have generally

FIGURE 3.2 Evolution of Concern for Ecosystems in the United States.

All Ecosystems	National Biological Diversity Conservation and Environmental Research Act
_____ 1990 _	_____
Rangeland	Federal Land Management and Policy Act (1976)
Forestlands	National Forest Management Act (1976)
Coastal Areas	Marine Protection Research and Sanctuaries Act (1972)
Rivers	Wild and Scenic Rivers Act (1964)
Wilderness	Wilderness Act (1964)
Wildlife Sanctuaries	Fish and Game Sanctuary Act (1934)
Grazing Lands	Taylor Grazing Act (1934)
_____ 1900 _	_____
Timberlands	Forest Reserve Act (1891)

resulted from increased awareness of threats to certain elements of biodiversity. The original game conservation laws in this country were a reaction to declining numbers of game species. Similarly, the original forest reserve legislation was a reaction to rapid decimation of public and private forests. Currently proposed legislation related to gene conservation is a reaction to an increased awareness of the value of genetic diversity (Schonewald–Cox et al. 1983).

The recent concern thus follows in this tradition. Loss of species and the ecosystems to which they belong is proceeding at an unprecedented rate. Since the threats to life are more imminent, more insidious, and more global than in the past, a more holistic concern and response has developed. But amidst all this concern over biological diversity in general, rangelands have received little attention in this country. Major attention has focused upon loss of biodiversity in forests, particularly on the liquidation of ancient forests. Yet biological diversity on rangelands is just as threatened. Because the threats are typically more subtle on rangelands, they receive less attention. Ehrlich (1989a) has pointed out that there was no reason for our ancestors to evolve a capacity to detect gradual environmental trends. Our rangelands similarly suffer from this myopia.

CURRENT THREATS

Biodiversity on western rangelands is threatened by many factors: diversion and pollution of water, agricultural and urban development, livestock grazing, mineral extraction, habitat fragmentation, and global warming. Water is in limited supply in the West, but it is in great demand for human activities. Thus some of the earliest and most pervasive effects on biological diversity have been due to diverting water from natural systems to human activities. Damming rivers, lowering water tables, degrading riparian areas, and usurping springs for livestock have had a major impact on the biota of the West (Hunt 1988).

The impact of livestock grazing on biological communities in the West has received much attention. Most of it, however, has focused on competition between livestock and large, huntable, or otherwise economically valuable animal species (Wagner 1978). Only in the last two decades have some of the less obvious effects of livestock grazing been noted: transmission of diseases from livestock to native species, as from domestic sheep to bighorn sheep (Goodsen 1983); decline of bird species due to loss of cover from livestock grazing (Brown 1978); and competition between livestock and smaller native herbivores such as desert tortoises (Berry 1978). The influence of livestock grazing on plants has been similarly widespread. But most of the attention has focused on the community level; very little is known, for example, about effects on rare plants. Furthermore, the impact of livestock grazing on invertebrates or lower plants is rarely considered.

These few examples represent only a small sample of the impact of introducing over 300 million animal unit months per year of livestock grazing onto western rangelands in a period of a little over 100 years.

Wagner (1978) has concluded that livestock grazing is certainly the most ubiquitous influence on the biota of the West. While some changes are drastic and obvious, others are subtle and not easily recognizable, even by professionals. Since livestock grazing remains one of the most common and widespread uses of western rangelands and its impacts on biological diversity are so poorly understood, such grazing must be considered one of the primary threats to biological diversity.

Other threats include the myriad activities on the arid lands of the West, particularly mining, agricultural development, recreation, urban/suburban expansion, and the development of transmission and transportation corridors. The threats to biological diversity from these activities are too numerous to document here, except to note that a common result is fragmentation of habitat. Splitting large blocks of continuous habitat into smaller isolated parcels that cannot support viable populations of plants or animals can cause local extirpation or eventually extinction (Wilcove 1987).

Finally, future global warming could result in substantially altered climatic regimes in the West, causing increased extinctions (Murphy 1988). Moreover, these changes are likely to be synergistic with other impacts (Peters 1988). For example, rangelands that are fragmented and heavily grazed, with deteriorated riparian areas and watersheds, are likely to suffer greater loss of biological diversity from increased temperatures and decreased precipitation.

CURRENT PROGRAMS

Blockstein (1989) has suggested that any program for conserving biological diversity should encompass four elements: designation and management of protected areas; utilization of seminatural areas; recovery of endangered species and degraded ecosystems; and research. What is the status of such efforts on western rangelands?

Designation and Management of Protected Areas. The goal here is to form a preserve system through the designation of representative samples of the ecosystems and communities of the West. This has been attempted in three ways. The oldest approach has been to designate at the largest scale national parks and monuments and certain larger wildlife and game refuges. The second way has been to designate wilderness areas and wild and scenic rivers within other primarily federal lands. Finally, at the smallest scale, protection has sometimes taken the form of natural areas, research natural areas, refuges, or areas of critical environmental concern (ACECs).

Current preserve systems in the United States are of limited effectiveness by themselves because most were not established to preserve biological diversity (Blockstein 1989). This point is well illustrated on western rangelands. Most of the national parks and monuments of the nonforested West—parks such as Grand Canyon, Zion, and Capitol Reef—were designated because of their spectacular geological features. Similarly, until recently most designation of wilderness areas and wild rivers has been based upon desirability for primitive recreation such as backpacking and canoeing. This policy has resulted in a disproportionate number of alpine wilderness areas and white-water wild and scenic rivers.

In the natural areas there is similar evidence. Williams and Campbell (1988) point out that even though 191 ACECs have been designated to protect natural values (including both plants and animals), the program has still been inconsistent. Nevertheless, positive signs of change are evident. The Bureau of Land Management (BLM) has systematically compared wilderness study areas (areas being studied for inclusion in a national wilderness system) to determine if they represent Bailey-Kuchler vegetation types (one system of classifying potential natural vegetation) not currently designated as wilderness. At latest count, BLM wilderness study areas represent 111 of 138 Bailey-Kuchler types in the West, many of which are not now in any wilderness system (Cooperrider 1989). Furthermore, BLM is conducting a parallel study to determine the representation of Bailey-Kuchler types within ACECs.

A second major problem with the preserve system in the West is that many preserves are not large enough to maintain viable populations of target species, much less self-sustaining ecosystems. Our oldest and largest national park in the West, Yellowstone, is not large enough to contain viable populations of many species, thus necessitating the need for management based on the "Greater Yellowstone Ecosystem" (Clark and Zaunbrecher 1987).

Finally, no preserve is truly pristine or totally protected. Air pollution, exotic plants and animals, polluted water, and other "nonnatural" elements cross preserve boundaries as readily as they cross county lines. Furthermore, global warming as well as other forms of global change will affect all areas equally, resulting in what has gloomily been termed "the end of nature" (McKibbin 1989).

Because of the limitations of preserves—and the unlikelihood that many new large preserves will be designated—management of the lands surrounding the preserves, the seminatural areas, is becoming of increasing importance (Salwasser 1987; Thomas and Salwasser 1989).

Management of Seminatural Areas. On western rangelands, some of the greatest opportunities for conserving biological diversity lie in the management of seminatural areas. The relatively recent settlement of the West and the high percentage of lands still in public ownership have resulted in a large amount of public land with limited loss of biological diversity (Cooperrider 1989). Of particular importance in western rangeland areas are the BLM lands that provide biological continuity to what would otherwise be isolated parks and refuges. Until quite recently, however, conservation of biological diversity on seminatural rangelands of the West, both public and private, has been accorded a lower priority than commodity production.

Ecological Recovery. Ecological recovery efforts are under way at both the species level and community level. At the species level, many efforts in the West are adequately publicized elsewhere. But here, as elsewhere in the world, the species approach to conserving biological diversity in the absence of habitat conservation is likely to fail (Hutto et al. 1987). Restoration at the community level, despite the widespread deterioration of many western rangelands, has been limited in scope and narrow in purpose. For example, the principal purpose of most rangeland rehabilitation projects has been restoration of livestock forage. Such projects typically end up reducing plant and animal species diversity.

A major exception to this pattern is the growing effort among resource agencies to rehabilitate riparian areas in the West. Riparian areas in general are among the richest in biological diversity, and in the arid West they are understandably overused by thirsty livestock in search of relief from the searing sun. The consequent degradation of these sensitive areas by livestock has led to some successful restoration efforts (Prichard and Upham 1986). Few other plant communities have received such attention, much less successful rehabilitation.

Research. Blockstein (1989) has emphasized the need for scientific research to address questions related to biodiversity. Although much research has been conducted on rangelands, until recently it has focused mainly on management for livestock production. Research relevant to conservation of biological diversity on rangelands has been quite limited (and most advances in our knowledge have been fortuitous). For example, much money has been spent on research into ways to control or exterminate native shrubs and trees such as sagebrush, juniper, and piñon pine, species alleged to limit production of livestock forage. Yet very little research effort has focused on the ecology of these plants and the biotic communities of which they are keystone species. Similarly, research on ways to control exotic plants has typically focused on high-technology treatments such as herbicides rather than on the ecology of plant invasions.

Closely related to the need for research is the need for inventory and monitoring. The biological resources of most rangeland areas of the West have never been inventoried. Furthermore, much useful information on rangeland management, including successful efforts to conserve biological diversity, has been lost through lack of adequate documentation.

NEW APPROACHES

Considering the many threats cited here, the risk of catastrophic losses of biodiversity on western rangelands is great. Many rangelands of the Great Basin are already showing symptoms in the form of widespread invasion of exotic plants. Desertification, or serious degradation of rangelands, is not limited to developing countries (Savory 1986).

Yet here is an opportunity to establish an exemplary program for conservation of biological diversity. Traditional approaches to conservation will need to be continued and even strengthened. There are still many areas in the West that need designation and protection as parks, wilderness, natural areas, or ACECs. Similarly, endangered species programs need more support as do land rehabilitation efforts. Furthermore, the longstanding need for an inventory of the biological resources of western rangelands will be a prerequisite to many programs.

More important, new approaches to conservation, involving both the government and the private sector, are needed. Such approaches to

conservation biology in the West should have the following characteristics: recognition of the complexity of ecosystems and ecosystem
processes; recognition of the need for planning and management at the
ecosystem or regional level (to complement other levels); use of a
variety of disciplines in studying and solving problems; and an unprecedented level of coordination between public and private agencies and
other interested parties.

Many human-caused losses of biological diversity have been the
result of simplistic notions of ecosystems and ecosystem processes. To
reverse this trend we must recognize the complexity of ecosystems
even though our knowledge of how they work may be very slow in
coming and will always be incomplete. Recognition of this complexity
will reinforce the need for interdisciplinary approaches to resource
problems. Appreciation of the complexity of ecosystems will similarly
discourage the use of quick-fix, high-tech solutions such as the use of
herbicides without knowledge of their long-term impacts.

Recognition of the complexity of rangeland ecosystems leads to the
need for research into basic ecosystem processes. Current research
suggests that ecosystems may be more easily upset by human perturbations than previously thought (Perry et al. 1989) and that virtually
unstudied elements such as mycorrhizal fungi may be keystone species
on rangelands (Trappe 1981). There has been much research on how to
maintain a few "desirable" forage species and how to eradicate certain
exotic or weedy "undesirable" species. Until fairly recently, however,
little research has focused on the effects of grazing on rare plants or,
more generally, how to maintain species diversity on grazed rangelands. Although a few recent studies have investigated the impact of
livestock grazing on riparian plant communities and the fauna they
support, the impact of grazing on plant species or plant genetics is
virtually unknown.

We need to support more research on how to maintain biological
diversity on rangelands. In conjunction with this effort, we need a
program of systematic inventory and monitoring of this biodiversity.
Finally, we need to take advantage of local ways of managing areas
with minimal impact on biological diversity (Dasmann 1985). This is a
concept often mentioned in relation to Third World countries but
typically neglected in our own backyard.

Recognition of the complexity of ecosystems reinforces the need for
planning at the ecosystem, landscape, or even regional level. Since

many ecosystem processes such as disturbance cycles and succession occur at these higher levels, planning for conservation of biological diversity must be done on this scale as well as others (Noss 1983). As pointed out earlier, most preserves in the West are too small to conserve biological diversity. Therefore, they must be managed in conjunction with surrounding seminatural lands, an effort that will require planning at the regional level. The gap analysis being pioneered in Idaho represents the sort of regional or landscape level of inventory and planning that is needed (Scott et al. 1987).

Effective programs on seminatural areas will have to be managed with the dual objectives of biodiversity conservation and resource production. Equally important, management programs for these areas must be developed in conjunction with programs for the embedded preserves. This concept is central to the notion of "multiple-use modules" proposed by Noss and Harris (1986) or the idea of biosphere reserves (Hough 1988). At present, such planning and management are rare. Preserves are often managed as if they existed in isolation. Surrounding seminatural lands are exploited for resource production at the expense of the substantial natural diversity they harbor.

Livestock grazing is the most widespread use of seminatural rangelands. The debate over the limits of tolerance to livestock grazing on western rangelands is highly polarized—some advocating removal of all livestock from all public lands, others minimizing the impacts or even suggesting the benefits to biological diversity from livestock grazing. Arguments abound regarding the impact of numbers, class of livestock, distribution, and timing of grazing.

The criterion of sustainability should provide the guidance needed to resolve this question. If livestock can be grazed in such a way as to preserve the biological diversity and ecological integrity of the landscape, there are few reasons to oppose such grazing. Some areas have been grazed by livestock for years with no apparent or measured loss of biological diversity. For many areas of the West, however, the premise that livestock can be grazed without loss of biological diversity remains untested. In other areas, livestock grazing is not compatible with maintaining biological diversity—for example, domestic sheep grazing within bighorn sheep range results in catastrophic dieoffs of the bighorn (Goodsen 1983).

Since landscapes and ecosystems typically comprise lands managed by a diversity of owners, implementation of programs will require unprecedented cooperation between public and private agencies. Managers of preserves must shed elitist attitudes that their pristine islands are not connected to the rest of the world and can be managed in isolation. Similarly, managers and users of seminatural rangelands must shed their commodity-oriented biases and recognize that maintenance of a landscape and its biological diversity is the foundation for continued resource production. Finally, private individuals and organizations must recognize that both the preserves and the surrounding areas are important, physically and spiritually, to their health and welfare.

Because of the vast amount of public, primarily federal, land in the West, the leadership role of the federal government is of prime importance. But federal leadership will not be enough. History has shown that programs lacking the support of local people are seldom successful in the long run.

THE IMPORTANCE OF AWARENESS

Thus we are led to the central importance of increased awareness of biological diversity. Genes, species, ecosystems, and the processes that maintain them are the basic components of sustainable systems. As these components are lost, ecosystems begin to unravel and eventually cease to function. Sustainable ecosystems are required for sustainable economies in spite of the blind economists who describe economic cycles with no external inputs—the economist's equivalent of the perpetual motion machine (Ehrlich 1989b). Thus, sustainable ecosystems are the basis of sustainable societies. Given this situation, preservation of biological diversity on rangelands is just as important to the rancher as to the birdwatcher or hunter.

A successful program for conserving biological diversity must recognize this relationship between biological diversity and sustainability. It will require awareness by all parties—politicians, bureaucrats, ranchers, landowners, conservationists, and citizens. And it will require action.

REFERENCES

Berry, K. H. 1978. Livestock grazing and the desert tortoise. *Trans. N. Am. Wildl. and Nat. Resources Conf.* 43:505–519.

Blockstein, D. E. 1988. U.S. legislative progress toward conserving biological diversity. *Conserv. Biol.* 2(4):311–313.

_____. 1989. Toward a federal plan for biological diversity. *Issues in Science and Technology* 5(4):63–67.

Brown, D. 1978. Grazing, grassland cover and gamebirds. *Trans. N. Am. Wildl. and Nat. Resources Conf.* 43:477–485.

Capen, D. E. 1989. Political unrest, progressive research, and professional education. *Wildl. Soc. Bull.* 17(4):335–337.

Clark, T. W., and D. Zaunbrecher. 1987. The greater Yellowstone ecosystem: the ecosystem concept in natural resource policy and management. *Renewable Resources J.* 5(3):8–16.

Cooperrider, A. Y. 1989. Conservation of biological diversity in the west. Pp. 3–9 in *Proceedings IV: issues and technology in the management of impacted wildlife.* Boulder: Thorne Ecological Institute.

Dasmann, R. F. 1985. Achieving the sustainable use of species and ecosystems. *Landscape Planning* 12(3):211–219.

Ehrlich, P. R. 1987. Habitats in crisis. *Wilderness* 50(176):12–15.

_____. 1989a. Facing the habitability crisis. *Bioscience* 39(7):480–482.

_____. 1989b. The limits to substitution: meta-resource depletion and a new economic-ecological paradigm. *Ecol. Econ.* 1(1):9–16.

Goodsen, N. 1983. Effects of domestic sheep grazing on bighorn sheep populations: a review. *Trans. Biennial Symposium of the Northern Wild Sheep and Goat Council* 3:287–313.

Hough, J. 1988. Biosphere reserves: myth and reality. *Endangered Species Update* 6(1–2):1–4.

Hunt, C. E. 1988. *Down by the river.* Washington: Island Press.

Hutto, R. L., S. Reel, and P. B. Landres. 1987. A critical evaluation of the species approach to biological conservation. *Endangered Species Update* 4(12):1–4.

Jacobs, P. 1986. Sustaining landscapes: sustaining societies. *Landscape and Urban Planning* 13:349–358.

McKibbin, B. 1989. The end of nature. *New Yorker*, Sept. 11, 1989, pp. 47–105.

Maser, C. 1988. *The redesigned forest.* San Pedro, Calif.: R. & E. Miles.

Murphy, D. D., and S. B. Weiss. 1988. The effects of climate change on regional biodiversity in the western United States: species losses and mechanisms. Paper presented at World Wildlife Fund conference on Consequences of the Greenhouse Effect for Biological Diversity, Oct. 4, 1988, Washington.

Noss, R. F. 1983. A regional landscape approach to maintain diversity. *Bioscience* 33(11):700–706.

Noss, R. F., and L. D. Harris. 1986. Nodes, networks, and MUMs:

preserving diversity at all scales. *Environ. Management* 10(3):299–309.

Perry, D. A., M. P. Amaranthus, J. G. Borchers, S. L. Borchers, and R. E. Brainerd. 1989. Bootstrapping in ecosystems. *Bioscience* 39(4):230–237.

Peters, R. L. 1988. Effects of global warming on species and habitats—an overview. *Endangered Species Update* 5(7):1–8.

Prichard, D., and L. Upham. 1986. Texas Creek riparian enhancement study. *Trans. N. Am. Wildl. and Nat. Resources Conf.* 51:298–303.

Salwasser, H. 1987. Editorial. *Conserv. Biol.* 1(4):275–277.

Savory, A. 1986. A solution to desertification and associated threats to wildlife and man. *Trans. N. Am. Wildl. and Nat. Resources Conf.* 51:116–124.

Schonewald-Cox, C. M., S. M. Chambers, F. MacBryde, and L. Thomas (eds.). 1983. *Genetics and conservation: a reference for managing wild animal and plant populations.* Menlo Park, Calif.: Benjamin/Cummings.

Scott, J. M., B. Csuti, J. D. Jacobi, and J. E. Estes. 1987. Species richness. *Bioscience* 37(11):782–788.

Soulé, M. E. (ed.). 1986. *Conservation biology: the science of scarcity and diversity.* Sunderland, Mass.: Sinauer.

Soulé, M. E., and B. A. Wilcox (eds.). 1980. *Conservation biology: an evolutionary-ecological perspective.* Sunderland, Mass.: Sinauer.

Teer, J. G. 1989. Conservation biology—a book review. *Wildl. Soc. Bull.* 17(4):337–339.

Thomas, J. W., and H. Salwasser. 1989. Bringing conservation biology into a position of influence in natural resource management. *Conserv. Biol.* 3(2):123–127.

Trappe, J. M. 1981. Mycorrhizae and productivity of arid and semiarid rangelands. Pp. 581–599 in *Advances in food producing systems for arid and semi arid lands.* New York: Academic Press.

U.S. Congress, Office of Technology Assessment. 1987. Technologies to maintain biological diversity. Report OTA-F-330. Washington: Government Printing Office.

Wagner, F. H. 1978. Livestock grazing and the livestock industry. Pp. 121–145 in H. P. Brokaw (ed.), *Wildlife and America.* Washington: Council on Environmental Quality.

Wilcove, D. S. 1987. From fragmentation to extinction. *Nat. Areas J.* 7(1):23–29.

Williams, D. C., and F. Campbell. 1988. How the Bureau of Land Management designates and protects Areas of Critical Environmental Concern: a status report, with a critical review by the Natural Resources Defense Council. *Nat. Areas J.* 8(4):231–237.

Wilson, E. O., and F. M. Peter (eds.). 1988. *Biodiversity.* Washington: National Academy Press.

4 Roles and Approaches of the USDA Forest Service

Hal Salwasser

BIOLOGICAL DIVERSITY is the variety of life and its many processes. It includes all lifeforms from bacteria to birds and mammals—possibly as many as 30 million different species worldwide (Wilson 1988) plus countless millions of subspecies and ecological processes that link organisms to communities, ecosystems, and, ultimately, the entire biosphere. (See Table 4.1.)

TABLE 4.1. **Number of Described Species of Living Organisms**

Kind of Organisms	Estimated Number of Described Species
Viruses	1,000
Monera	4,760
Fungi	46,983
Algae	26,900
Higher plants	248,428
Protozoa	30,800
Invertebrates	989,761
Chordates	43,853
Total	1,392,485[a]

Source: Derived from Wilson (1988: table 1-1, pp. 4–5).

[a] Our knowledge of the true number of species of living organisms is very incomplete. Wilson, an entomologist, believes the true number of invertebrates to exceed 5 million and suspects that the absolute number of all species falls somewhere between 5 and 30 million.

It would be nice if we could restore or perpetuate every part and process of a diverse biota in every possible place. But a growing human population and its corresponding stresses on our global environment mean that we must continually make choices on conservation priorities and blend those choices with other social goals. For biodiversity, this means we must often select options that only reduce the rate of loss or only approximate desired ecological conditions (Cairns 1989).

I base this discussion on three assumptions (the merits of which can be debated elsewhere): Biodiversity is a basic resource for all life; no single entity—public or private—can perpetuate biodiversity by itself; and federal land and resource management agencies must play a pivotal role in an overall national conservation strategy.

VALUES OF BIOLOGICAL DIVERSITY

It is axiomatic that we cannot perpetuate all biological diversity while extracting an increasing portion of its productivity for growing human uses. Nor can we perpetuate desired species, communities, and ecological processes simply by trying to stop human progress or by preserving a few isolated natural areas and trying to prevent change in them. Though the idea is initially appealing, we have also learned that trying to maximize the variety of life on every acre or in every watershed is not the solution (Samson and Knopf 1982).

What should we care about first? And why is that more important than other concerns? Regardless of the ecological notion that all species are of equal value, the answer to these questions is not likely to derive solely from scientific principles. It will be greatly influenced by economics, aesthetics, and ethics, as well as by biology and ecology.

Perhaps a look at how the variety of life serves people will offer some insight into immediate priorities, though this is not meant to imply that utilitarian values are the sole, or even primary, reasons to conserve biological diversity. Diverse communities of plants, animals, and microorganisms provide indispensable ecological services. They recycle wastes, maintain the chemical composition of the atmosphere, and play a major role in determining the world's climate. The biota also contains the genetic resources for crops and medicines.

As great as the realized values are, the potential values and uses of biodiversity far exceed our current knowledge (Oldfield 1984). We know only a small fraction of the species on this planet, especially in tropical ecosystems, despite decades of scientific effort (Lugo 1988). Every year species are lost before we have a chance to describe them, let alone learn anything of their ecological roles and values (Ehrlich 1988; Myers 1988). We will never know which potential foods, medicines, and commercial products are lost with each reduction in biological diversity.

PERSPECTIVES ON A CONSERVATION STRATEGY

With all this variety of life, its values, and its vulnerability to human actions, how do we move forward on a conservation strategy? I will mention seven items I think are important to an overall strategy. There are probably others as well.

First, we must recognize that conservation of biological diversity requires a "big picture" perspective that recognizes humans and change as fundamental to any solution. Management of genetic resources, species populations, ecosystems, bioregions, and human activities to perpetuate the variety of life while meeting short-term and long-term human needs must integrate many considerations. Such management must also employ a full spectrum of conservation actions—from protection, restoration, enhancement, and sustainable culture to research, inventory, assessment, planning, monitoring, interpretation, education, and public relations. Conservationists have been engaged in such efforts for many decades. But they have tended to focus on specific resources and specific places, seldom with biodiversity as an overall concern.

Addressing specific resources—such as endangered or featured species or specified places such as Yosemite National Park—as if they were ecologically isolated rarely requires attention to another "big picture" perspective: regional and continental geographic scales. Attention to biodiversity and the global environmental changes resulting from human activities mandates universal consideration of these larger geographic scales.

Second, to approach such a comprehensive goal as perpetuating biodiversity we need a better understanding of the structure, functions, and processes of regional populations and ecosystems. We must also understand their responses to the cumulative effects of environmental change, human activities, and resource management. Then we must make prudent decisions on how to rehabilitate, sustain, or enhance ecosystem productivity for all values and uses. This means that we must begin a dialogue and experimental actions to diversify our theories and methods for carrying out conservation of lands and resources.

Third, biodiversity is a political issue because it portends a change in social priorities and will probably entail certain changes in current conservation practices. Without a compelling recognition of the need for action, making those changes will be difficult. We must set the course for a biologically diverse world before all the science is done (Soulé 1986; Cairns 1988b). Thus we must begin with a national commitment: a focus and sense of urgency to mobilize attention.

Many federal and state agencies, together with private entities, are addressing the conservation of biological diversity. But, overall, the policies and actions for conserving biodiversity in the United States are fragmented. We need to integrate them into a comprehensive conservation strategy. This may or may not require new legislative mandates; much can be done by building from existing laws, regulations, policies, and institutions. We have yet to take full advantage of established agency mandates and capabilities.

A national strategy should be coordinated between all the agencies and organizations—public and private—that share interests or responsibilities for natural resources (U.S. Congress 1987). We must reduce the barriers—real and perceived—between functional disciplines such as timber and wildlife, science and management, parks and multiple-use forests, federal and state agencies, and government and the private sector. On the issue of biodiversity, they must all play complementary roles.

Fourth, many of the rarest elements of biodiversity occur on public lands—some are indigenous to these lands, others find their last refuge there. We develop and change policies for how public lands and resources are managed by democratic means. Our society used to delegate nearly complete authority for management of public resources to professionally trained rangers, biologists, game managers, and foresters. Those days are gone. We have an increasingly knowledgeable and concerned citizenry that wants a voice in making the tough choices regarding resources.

Agencies must open their planning and resource management processes even more than they have done recently. The alternatives—resolving controversy through litigation or smoke-filled rooms; or increasingly centralized, technology-driven planning—waste time and talent and run counter to our democratic traditions of solving problems through consensus.

Fifth, regional perspectives and integrated ecosystem management require new technologies. Research and technology development must embrace large-scale, long-term biodiversity issues such as population viability, ecological resilience, landscape linkages, ecological restoration, multiresource yield functions, habitat fragmentation, biodiversity indicators, cumulative effects analysis, and monitoring. There is an implication here for both education and science: They must become more, not less, interdisciplinary, and both must increasingly focus on adaptive approaches to solving problems that defy technical solutions. Research and development must become integral parts of management strategies and their implementation.

Sixth, public relations is also needed. Most people know what spruce trees and bald eagles are; some even know the roles of management in their conservation. But biodiversity is another story. Scientists have yet to reach a consensus on what it is, and hardly anyone is willing to admit they know how to conserve it while meeting other human needs. Thus we have an enormous job ahead of interpretation, education, and persuasion to gain awareness, understanding, and support.

Finally, the keys to a successful strategy to conserve biodiversity are integration and on-the-ground action. We must take advantage of all our lands and resources, from the most highly protected to the most intensively managed, and use all the management tools we can muster.

Conservation of biodiversity can start with reserved areas such as parks, wilderness, and natural areas, but it cannot stop there. Only about 7.5 percent of the United States is in parks, wilderness, and nature preserves, and even then we are ahead of other nations on such land protection. Multiple-use lands will be critical to the solution (Figure 4.1). Some have argued (Norse et al. 1986; Wilcove 1988) that they are even more important to the solution than parks and preserves. But even they will not be enough. We cannot meet our resource production needs and save major elements of biodiversity just on public, multiple-use, and reserved lands. Private lands, including

FIGURE 4.1 Land Uses in the Continental United States
(Millions of Acres).

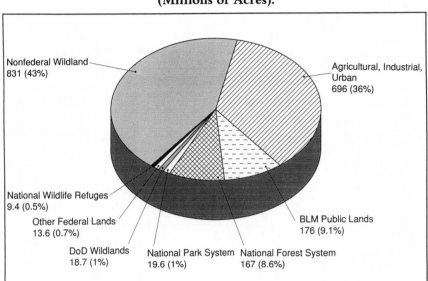

Nonfederal Wildland
831 (43%)

Agricultural, Industrial,
Urban
696 (36%)

National Wildlife Refuges
9.4 (0.5%)

Other Federal Lands
13.6 (0.7%)

DoD Wildlands
18.7 (1%)

National Park System
19.6 (1%)

National Forest System
167 (8.6%)

BLM Public Lands
176 (9.1%)

Data from World Resources Institute et al. (1988) and USDI land statistics.

those benignly neglected and those managed for products, must play
significant roles also. High productivity from any piece of land or
water will allow other parcels to be managed for other purposes,
including restoration or enhancement of some rare or sensitive element
of biodiversity.

A comprehensive, long-term regional perspective on biodiversity
implies that we will integrate our plans and actions along three dimensions: biological, temporal, and geographic. We can get a feel for the
need to integrate different dimensions by thinking of it as blending our
plans and actions for focal components of diversity—from stands to
watersheds to regions—with attention to consequences for long periods. The need for increased integration is fundamental to biodiversity conservation, and it is a far greater challenge than anything
conservation science or management have yet tackled.

ROLES FOR THE FOREST SERVICE

The USDA Forest Service can play an important role in an overall national strategy for biodiversity. We can start with stewardship of the 191-million-acre National Forest System—8.6 percent of the nation's total area. The agency can also contribute to the development of national and international policies, basic knowledge, and technologies. And it can help interpret, educate, and demonstrate the need for sustainable management of renewable natural resources.

A beginning approach is being developed for national forests. Consistent with scientific knowledge and guided by the legal mandates of the Forest Service's mission, the approach is designed to provide sustainable yields of all the multiple uses and values of forests. "Providing for diversity of plant and animal communities to meet overall multiple-use objectives" is the legal mandate of the National Forest Management Act of 1976. It sets a new focus for programs, plans, and actions to achieve specific results for biota in the National Forest System. The evolving focus on biodiversity now includes several elements: recovering threatened and endangered species to viable levels; managing habitats and human actions to maintain viability of all other species' populations well distributed throughout their geographic ranges; protecting, managing, restoring, or enhancing special habitats and a network of natural biological communities; ensuring the structural and functional integrity of regional ecosystems; and managing the genetics and populations of species desired for human well-being. Forest Service programs and plans for diversity vary as we are still learning how to approach this complex issue.

Like most agencies, the Forest Service needs better inventories of the basic elements of biodiversity. In the National Forest System, these elements are probably known better than for any other large land-management system, though they are still known only generally for our comprehensive management purposes. For example, the national forests and national grasslands contain nearly 80 percent of the vertebrate species richness in the United States, 30 percent of the listed endangered and threatened species, over 70 percent of the major vegetation types, 50 percent of the salmon and trout habitat, 50 percent of the standing volume of sawtimber, and 75 percent of the big-game populations in the West. These areas also form the backbone of some large, contiguous wildland ecosystems that range from the 3-million-acre Southern Appalachian Highlands to the 32-million-acre Northern Continental Divide (Figure 4.2). And this only scratches the surface of the base for biodiversity on these lands. We need better inventories and assessments of the current conditions; we need better abundances and distributions; and we need better management direction for genetic resources, species populations, biological communities, and ecological systems. Natural Heritage programs (Jenkins 1988) are perhaps the best example of coordinated inventories, and Geographic Information Systems are the most promising of new technologies (Scott et al. 1987).

Yet even a system as large and diverse as a national forest cannot sustain or rehabilitate its biodiversity without cooperation of adjoining landowners and managers. We need to bring common inventories, research, assessments, plans, and policies into better coordination and coverage. Such multi-institutional coordination of plans and actions will be difficult. Only in Yellowstone, the Southern Appalachian Highlands, and the Columbia River fisheries have we approached this task. People must yield some of their autonomy and there will, no doubt, be adjustments in how areas of lands and waters are managed. That probably means some changes in public policies for multi-institutional resource conservation.

Each type of landownership can do some things better than others. What specifically can the national forests and national grasslands do better or differently for biodiversity? For starters, they contain over 32 million acres of federally designated wilderness—an area larger than the entire National Park System in the lower forty-eight states. They also hold much of the last remaining habitats for over 170 federally listed threatened or endangered species. Their recovery would remove 30 percent of these species from the federal list. The habitats of nearly 1,000 sensitive or candidate species also occur in the National Forest System. Securing their viability in well-distributed populations would go a long way toward perpetuating forest diversity in the United States. National forests are an unequaled storehouse of genetic diversity, especially for species of high commercial value. There is no valid reason to prolong action on the vital role that management of genetic resources in national forests can play (Ledig 1986).

The national forests also hold some of the most highly valued

FIGURE 4.2 Potential Bioregions Within the National Forest System.

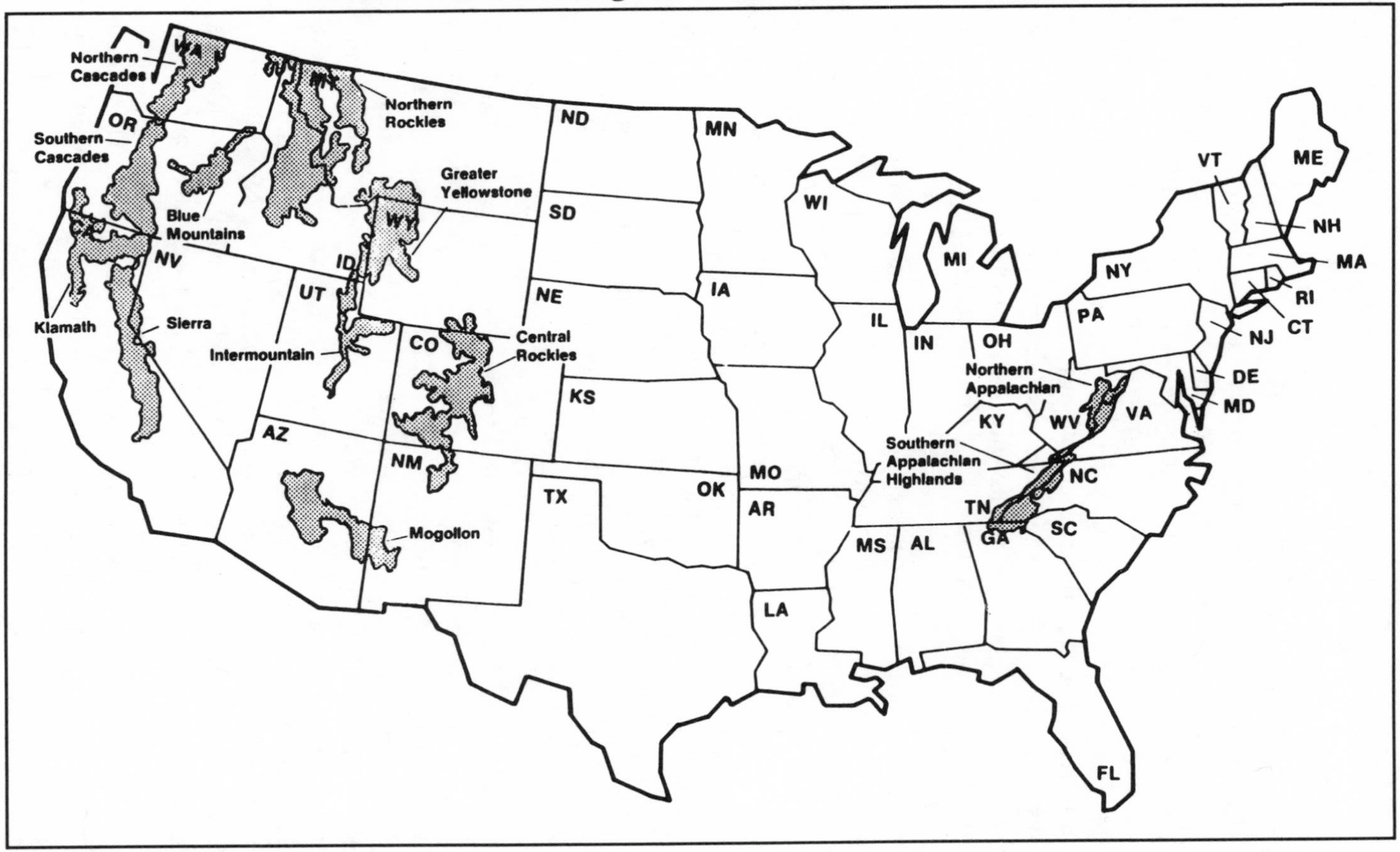

Adapted from Salwasser et al. (1987).

biological communities in the country. These include most of the remaining old-growth forests, over a hundred thousand miles of riparian areas, virtually all of the alpine ecosystems that occur outside national parks, and potentially the most complete system of research natural areas in the country—currently more than 200 areas covering over 200,000 acres. The national forests contain some of the nation's highest-valued resources. This includes 57 million acres of commercial timberland and such world-class resources as wild turkey, black walnut, elk, Douglas fir, bighorn sheep, coho salmon, and native trout.

Finally, the Forest Service is pioneering in opening the public resource decision-making process to broad involvement and interagency coordination. Much controversy exists over how public forests are to be managed, but that is the cost of change. And change is essential for conservation to be responsive to shifting market forces, new attitudes, and improved knowledge.

A WORTHWHILE JOURNEY

Conserving biological diversity is an ideal. It may be only partially achievable. But we should aim far beyond what we think we can achieve. As the poet Goethe said: "Dream bold dreams; they're full of power and magic." Conserving biodiversity while meeting people's needs for resources is not going to be easy. Goals for specific ecological conditions will have to compete with goals for specific resources. Trade-offs will be inevitable because there is only one land base and there is, as they say, no free lunch.

Conservation of resources for multiple values and multiple uses has come a long way on the path toward biodiversity in recent years. But we also have a long way to go. The journey is certain to be challenging, difficult, and not without a few dead ends. Yet it is a worthwhile journey. The future richness and productivity of life on earth depend on our taking it. And the future quality of life for all life depends on our reaching the destination.

REFERENCES

Cairns, J. Jr. (ed.). 1988a. *Rehabilitating damaged ecosystems.* 2 vols. Boca Raton, Fla.: CRC Press.

——————. 1988b. Restoration ecology: the new frontier. In J. Cairns, Jr. (ed.), *Rehabilitating damaged ecosystems,* vol. 1. Boca Raton, Fla.: CRC Press.

——————. 1989. Restoring damaged ecosystems: is predisturbance condition a viable option? *Environ. Prof.* 11:152–159.

Ehrlich, P. R. 1988. The loss of diversity: causes and consequences. In E. O. Wilson (ed.), *Biodiversity.* Washington: National Academy Press.

Holling, C. S. (ed.). 1978. *Adaptive environmental assessment and management.* New York: Wiley.

Jenkins, R. E. Jr. 1988. Information management for the conservation of biodiversity. In E. O. Wilson (ed.), *Biodiversity.* Washington: National Academy Press.

Ledig, F. T. 1986. Heterozygosity, heterosis, and fitness in outbreeding plants. In M. E. Soulé (ed.), *Conservation biology: the science of scarcity and diversity.* Sunderland, Mass.: Sinauer.

Lugo, A. E. 1988. Estimating reductions in the diversity of tropical forest species. In E. O. Wilson (ed.), *Biodiversity.* Washington: National Academy Press.

Myers, N. 1988. Tropical forests and their species: going, going . . . ? In E. O. Wilson (ed.), *Biodiversity.* Washington: National Academy Press.

Norse, E., K. Rosenbaum, D. Wilcove, B. Wilcox, W. Romme, D. Johnston, and M. Stout. 1986. *Conserving biological diversity in our national forests.* Washington: Wilderness Society.

Oldfield, M. L. 1984. *The value of conserving genetic resources.* Washington: USDI National Park Service.

Salwasser, H., C. M. Schonewald-Cox, and R. Baker. 1987. The role of interagency cooperation in managing for viable populations. In M. E. Soulé (ed.), *Viable populations for conservation.* Cambridge: Cambridge University Press.

Samson, F. B., and F. L. Knopf. 1982. In search of a diversity ethic for wildlife management. *Trans. 47th N. Am. Wildl. and Nat. Resources Conf.* 47:421–431.

Scott, J. M., B. Csuti, J. D. Jacobi, and J. E. Estes. 1987. Species richness: a geographic approach to protecting future biological diversity. *Bioscience* 37(11):782–788.

Soulé, M. E. 1986. Conservation biology and the real world. In M. E. Soulé (ed.), *Conservation biology: the science of scarcity and diversity.* Sunderland, Mass.: Sinauer.

U.S. Congress. 1987. Technologies to maintain biological diversity. OTA-F-331, GPO Stock no. 052-003-01058-5. Washington: Government Printing Office.

Walters, C. 1986. *Adaptive management of renewable resources.* New York: Macmillan.

Wilcove, D. S. 1988. *National forests: policies for the future.* Vol. 2: *Protecting biological diversity.* Washington: Wilderness Society.

Wilson, E. O. 1988. The current state of biodiversity. In E. O. Wilson (ed.), *Biodiversity.* Washington: National Academy Press.

Wondolleck, J. M. 1988. *Public lands conflict and resolution: managing national forest disputes.* New York: Plenum Press.

World Resources Institute, International Institute for Environment and Development, and UN Environment Programme. 1988. *World Resources 1988–89: An Assessment of the Resource Base That Supports the Global Economy.* New York: Basic Books.

5 Landscape Protection and The Nature Conservancy

Bennett A. Brown

THE NATURE CONSERVANCY is the only major national organization that is entirely devoted to the conservation of biological diversity. Traditionally, its methods have focused on the establishment of nature preserves, both in concert with government agencies and through additions to the Conservancy's own private preserve system—the largest such system in the world. The Conservancy's efforts involve three separate functions: the identification and design of nature reserves, the protection of land, and the long-term management (stewardship) of biological elements of natural diversity in protected areas.

To define its goals and set realistic priorities, the Conservancy has worked with state and federal agencies to establish a national network of programs to assess the status of biological diversity. The first such program was established in North Carolina in 1974. There are now Natural Heritage inventories in all fifty states and Conservation Data Centers (CDCs) in thirteen Latin American countries. In addition, regional data centers have been established in conjunction with the Navajo Nation and the Tennessee Valley Authority. Data centers have also been established cooperatively with the National Park Service to catalog biological diversity in the Great Smoky Mountains and the national parks of southern Florida.

In developing the Natural Heritage inventories, it was decided that a critical issue needed to be addressed: the loss of information stored in the genome of species that go extinct. Therefore, the first concern was

66

deemed to be the preservation of species diversity. To facilitate this effort, the Conservancy adopted a "coarse-filter/fine-filter" approach to cataloging and protecting biological diversity.

A COARSE-FILTER/FINE-FILTER APPROACH

The coarse filter of Natural Heritage inventories is the "natural community." Each program ecologist develops a classification of the natural communities known to occur in that jurisdiction. If one could find and protect intact examples of each natural community in an area, a large proportion of the area's species would be protected as well, since most exist as components (or at least inhabitants) of these natural communities. But there are species whose distributions do not coincide with those of natural communities or whose existence has been so disturbed by human activities or random events that they occur only in relict, localized, and often isolated populations. These are species whose continued existence cannot be assured through the protection of representative examples of natural communities. Thus the fine filter in Natural Heritage inventories is the "heritage lists" of plant and animal taxa that must be monitored and protected individually if we are to ensure their continued survival.

Using a ranking system to evaluate the rarity and overall endangerment of an "element of natural diversity" (that is, a species or a community), the Conservancy establishes priority lists of projects. It then goes about protecting these elements through the application of a variety of protection tools. As the Natural Heritage programs came into being during the late 1970s and early 1980s, however, most protection efforts involved the establishment of a series of preserves and managed areas. Since rarity and endangerment are primary criteria in establishing which elements are given a high priority, many of these initial efforts were aimed at stabilizing and protecting small, remnant examples of rare plants, animals, and natural communities. Often these remnants occurred in highly degraded, developed, or otherwise disturbed landscapes. The long-term viability of such preserves (and the elements they are supposed to protect) has often been questioned (Noss 1987).

Once an area has been protected, the Conservancy's policy is to determine who is most capable of providing effective long-term stewardship. This policy has resulted in many of the larger, more manageable tracts being transferred to local, state, or federal agencies. If a property is too small or the protected elements are so sensitive that they demand the strictest protection measures, it usually remains part of the Conservancy's preserve portfolio.

The protection activities described here are perhaps the best course for "lifeboating" extremely rare or threatened species or communities over the short term (Jenkins 1990). To ensure viability of protected elements over the longer term (and to preserve the integrity of evolutionary processes), more attention must be devoted to questions of scale in preserve design and management (Jenkins 1990; Baker 1989; Noss 1987). This has been done (with increasing frequency) over the last ten years in a number of projects. Preserves such as Santa Cruz Island off the coast of California, Pine Butte Swamp along Montana's Rocky Mountain front, Muleshoe Ranch and Aravaipa Canyon in Arizona, the Niobrara Valley Preserve in Nebraska, and the Virginia Coast Reserve are representative of landscape-scale protection projects the Conservancy has undertaken in the last decade and a half. As the ability to manage and fund large-scale projects has grown, there has been less hesitation to undertake landscape-scale protection efforts.

Although methods for identifying and designing landscape linkages and corridors are not as well defined, there have been efforts to protect these, as well, where it has been possible to clearly establish their importance. On the Amtrak Preserve in the Upper Midwest, a railroad right-of-way supports a corridor of intact tall-grass prairie plants. A conservation easement and a management agreement with Amtrak relieve the railroad of the need to treat its right-of-way with herbicides annually. The compact also allows the Conservancy to manage the right-of-way to maintain this swath of tall-grass prairie through an intensively developed landscape.

In the Sun Belt, a project by the Texas Nature Conservancy ensured the protection of an old-growth riparian corridor on the northwest fringe of Houston during the frantic real estate boom of the early 1980s. Cooperative projects with the U.S. Fish and Wildlife Service have ensured the protection of corridors for wildlife along the lower Rio Grande, both in Texas and adjacent Mexico. These borderland landscape linkages will help to prevent the genetic isolation of rare plants and animals in extreme south Texas.

Recently two developments have guaranteed that the Conservancy will increase its proportion of landscape-scale projects. First, by networking with cooperating state Natural Heritage inventories and Conservation Data Centers, the Conservancy is in a stronger position to identify and design such projects. The information generated by this network of data centers can be used to identify conservation needs across political boundaries. The network's classification of ecological communities is now almost completely regionalized, and soon these regional classifications will be merged into a national framework. As a result, we will have a broad perspective of the status of biological diversity in North America.

Second, the Conservancy's board of governors has recently given its preliminary approval to a new "bioreserve" protection campaign that will eventually target a number of landscape-scale sites in the United States and Latin America for protection. Evolved from the concept of the biosphere reserves of UNESCO's Man and the Biosphere program (UNESCO 1974; Franklin 1977), these projects seek to maintain ecological processes on an appropriate scale. Activities will be directed at landscape-scale projects in which a variety of protection tools can be used to ensure the integrity of large systems over time.

THE CAMPAIGN TO ESTABLISH BIORESERVES

To qualify, a potential bioreserve must consist of an ecologically viable landscape encompassing one or more of the following features:

- High-quality examples of terrestrial or aquatic communities that are endangered or inadequately protected
- Concentrations of rare species
- A large, relatively undisturbed example of a natural community once characteristic of its ecoregion but now fragmented or degraded
- A critical migratory stopover point or corridor

It is not the Conservancy's intention to acquire all the land in these bioreserve sites. Rather, the idea is to coordinate the use of a broad array of protection tools on the landscape. This will include protecting public and private uses of the land wherever appropriate. As Jenkins (1990) has pointed out, the concept of a preserve complex consisting of multiple ownerships and multiple zoning is not new: New York's Adirondacks

State Park was established in 1892; Britain's "greenline" system of national parks dates from the late 1940s. The Conservancy will make extensive use of traditional protection methods: conservation easements, management agreements, acquisition of key parcels, special designation of publicly owned lands, and cooperative acquisitions in behalf of various agencies. It will be necessary to deal creatively with such issues as zoning, land use, and comprehensive planning.

Bioreserves are intended to embrace one or more strictly protected "core areas" that are surrounded by buffer zones. Land use and resource management in the buffer zones are restricted to ensure that only reasonably compatible activities occur. As an example, the Conservancy has bought up strategically located agricultural tracts on the mainland of Virginia's Eastern Shore that are being resold with appropriate easements and deed restrictions. This policy permits compatible economic enterprises to function in these areas but prohibits those that may harm the long-term viability of the adjacent protected lands.

Cooperative efforts among Natural Heritage programs and Conservation Data Centers have resulted in the preliminary identification of approximately 300 potential bioreserve projects nationwide and in Latin America. The selection process will result in the eventual designation of no more than forty operational bioreserves. To deal more effectively with the planning of these bioreserve projects, information management will be improved with the addition of computerized mapping and Geographic Information System (GIS) technology to the Biological and Conservation Data System now used throughout the Natural Heritage network.

Unlike conventional protection projects that focus on rare or threatened elements, bioreserves also seek to preserve the "representativeness" of landscapes and ecoregions. Melding the technology of GIS and gradient analyses (Gauch 1982), Conservancy scientists are evaluating alternative sets of boundaries for selected bioreserves. These investigations will yield techniques to gauge the representativeness of different bioreserve configurations, quantitative measures of ecological diversity within alternative configurations, and estimates of the cost-effectiveness of alternative protection strategies.

Natural Heritage data centers have also been working with the U.S. Fish and Wildlife Service's Gap Analysis Project (Scott et al. 1987). The pilot project in Idaho has made extensive use of the Idaho Natural Heritage program's staff. Using information from the various Natural

Heritage databases, staff scientists produced the vegetation maps and vertebrate range maps for the Idaho gap analysis. Other Natural Heritage program staffs are cooperating in the production of vegetation and vertebrate range maps for the Pacific Northwest module. Moreover, the point location data for vertebrates contained in the Natural Heritage programs' "element occurrence" databases are being used to test the accuracy of the vertebrate range maps.

The cooperative development of accurate range maps and large-scale vegetation maps has enhanced our own ability to deal with landscape-scale protection projects. Collaboration with the Gap Analysis Project has permitted us to accelerate the regionalization of our community classification work in the western United States. Information derived from the gap analyses should enable not only the Conservancy but the entire conservation community to reevaluate our information needs and assist in setting research and protection priorities.

REFERENCES

Baker, W. L. 1989. Landscape ecology and nature reserve design in the Boundary Waters Canoe Area, Minnesota. *Ecology* 70(1):23–35.

Franklin, J. F. 1977. The biosphere reserve program in the United States. *Science* 195(1):262–267.

Gauch, H. G. Jr. 1982. *Multivariate analysis in community ecology.* Cambridge: Cambridge University Press.

Jenkins, R. E. 1990. Long-term conservation and preserve complexes. *Nature Conservancy Mag.* 39(1):4–7.

Keystone Foundation. 1990. Biological diversity on federal lands: a report of a keystone policy dialogue. Unpublished MS.

Noss, R. F. 1987. From plant communities to landscapes in conservation inventories: a look at The Nature Conservancy (USA). *Biol. Conserv.* 41:11–37.

Omernik, J. M. 1987. Ecoregions of the conterminous United States. *Ann. Assoc. Am. Geog.* 77:118–125.

Scott, J. M., B. Csuti, J. J. Jacobi, and J. E. Estes. 1987. Species richness: a geographic approach to protecting future biological diversity. *Bioscience* 37:782–788.

UNESCO. 1974. Programme on man and the biosphere (MAB) taskforce on: criteria and guidelines for the choice and establishment of biosphere reserves. MAB Report Series no. 22. Paris: UNESCO.

Discussion

The following discussion reflects the audience's questions concerning the broad issue of biodiversity conservation—what it means and how it can be effectively applied.

CRAIG MCWILLIAMS (TEXAS PARKS AND WILDLIFE DEPARTMENT): I would like to address my question to any member of the panel. Biodiversity and ecosystem management are what we're all talking about here. Is that understood to mean lake, sea, climate, species, or what?

REED NOSS: The way in which biodiversity is being discussed in conservation biology is very broad and certainly not limited to lake, sea, climate, or species. The goal is to maintain native biological diversity; genetic variety, both within and among species; the native composition and abundance of species in each regional landscape; the variety of community types of ecosystems that we see distributed across the land; and even the interconnections and natural patterns in which these ecosystems fall. Biodiversity is very broad, but I think we can identify measurable parameters or indicators that management can work with.

MICHAEL SOULÉ: If you're asking whether or not habitat management always considers successional stages, the answer is not necessarily so. For example, if a small patch of habitat is managed in such a way as to prohibit fire, then all the early successional organisms in that little patch would disappear, and some of those might be endangered plant species. So the proper frequency of disturbance to maintain a balance between successional stages is certainly an important aspect of biodiversity conservation. If fire frequency is too high, some late successional species might disappear. An example from southern Cali-

72

fornia, and one that is somewhat related to this, is the chaparral community. The rate at which chaparral is burned by land managers to maintain a low fuel load to protect people's homes is such that the species in chaparral will most likely go extinct in the near future. The fire frequency, dictated by policy, is about every ten to twenty years. That doesn't give enough time for the community to recover its diversity following a burn. So the smaller the patches of habitat we have to manage, the more intensively we have to manage them and the more attention has to be paid to succession and other aspects of ecosystem ecology.

BETTINA SPARROW (U.S. FISH AND WILDLIFE SERVICE): In our agency we've been having, as you may know, a bit of a battle between conservation biologists and wildlife biologists, so I appreciate that this workshop is attempting to bring both the theoretical and applied together. In terms of the applied, I'd like to know how do we really define biological diversity, and how do we manage the various types of biological diversity out there? It seems to me that if we can't answer these basic questions, then we can't go forward with one voice and educate the public.

MICHAEL SCOTT: I think the first step would be to manage as much as possible for self-sustained native ecosystems comprised of native species. In this manner, the entire array of species from the canopy down into the soil—species which, in some cases, might not yet have a scientific name—get managed. We are not going to protect them all that way, but I think it's the first step. What we really want to avoid is getting locked into managing single species along the lines of, say, the California condor or the black-footed ferret. We cannot avoid entirely this "emergency room" approach, but we would like to minimize the number of cases we have to deal with in that fashion. So I would say start at The Nature Conservancy's coarse level—the natural community level—and realize that big portions of overall biological diversity are common within those natural communities, even though you might not be able to protect viable portions of each and every natural community in the United States.

FELICE PACE: I would just like to add a few things. One is that you can't afford to wait until we have scientific certainty. Two, you know we need to act now because the loss of these habitats is progressing daily. Also, a fellow named Marc Liverman with the Portland Audubon Society is pursuing important legislation on an Endangered Ecosystem Act that would go beyond the Endangered Species Act. There are also things that we can do. Connectivity is something that seems to be a natural pattern. Maintaining connections or reconnecting fragmented areas are things I think people can respond to intuitively—and, therefore, don't need a lot of educating. I think we just need to get the word out and people will respond to it.

HAL SALWASSER: I'd like to add just one thing that is sadly a bitter reality. There's no question that protection of various habitats is a critical component in any strategy. But the reality is we will never protect enough big areas to represent all the variety of life that we might wish to sustain. And if we cannot develop some approaches to integrating biological diversity in managed landscapes while meeting people's needs, we are not going to get there.

MICHAEL SCOTT: I'd like to add that if we decide that natural vegetative communities are a reasonable surrogate for biodiversity, and species richness is another reasonable surrogate, we can take that first survey by using gap analysis and by taking advantage of the vast resources of The Nature Conservancy's Natural Heritage databases. It won't be the definitive survey, and it's not the biological inventory that's been called for. But we can do a gap analysis inventory of species richness and natural vegetative communities throughout the United States within six years and for less than one penny an acre.

GARY BARRETT: We've been talking about the importance of scale. We should also recognize that when ecologists increase scale from the ecosystem to the landscape level, we must also scale up our integrative research and funding efforts, as well. For example, the National Science Foundation, under its LTER (Long-Term Ecological Research) Program, has seventeen LTER sites presently being funded within the Ecosystem Studies Program but still has not integrated the socioeconomic components with the ecosystem component at the landscape level. In other words, we don't have, in my opinion, a robust landscape LTER site in operation. It will be interesting during the 1990s to see if a new and truly integrative LTER site in landscape ecology is funded either by NSF or as an interagency endeavor. This is going to be a challenge because it will demand a new cooperative approach to integrative research at a larger temporal/spatial scale and will demand a new research perspective. In time, this approach must integrate human-built systems (such as cities) or subsidized systems (such as forests) at the landscape level.

LARRY HARRIS: I can't speak for the panel, but I can tell you that I think there's a hell of a lot of sandbagging going on for several reasons. Can you define water? No. Can you define clear water? I doubt it. Can you define dirty water? The same can be asked for soil, air, and most of our resources. It's silly to listen to these repetitive harangues about how you can't do anything about something because you can't define it.

MICHAEL SOULÉ: I'd like to add that I find it shocking that we are still trying to define biological diversity after all the efforts of the Office of Technology Assessment and E. O. Wilson's book, *Biodiversity*, published by the National Academy of Sciences. That aside, I think there's a useful analogy to draw here. Hundreds of language groups have disappeared from the United States over the last 500 years. Mostly Native American groups. The same thing is happening in South America right now, where only about 10 percent, I recall, of the language groups remain. We are seeing an increasingly homogeneous world where everyone wears the same kind of designer jeans and digital watches and speaks the same language. There's a loss of human diversity due to television and consumerism and fax machines. . . . I think what Bettina was getting at was how do we communicate the seriousness of the loss of biological (instead of cultural) diversity to the people who make decisions? I think maybe use of this human analogy is one way. Another is simply to ask the following question: How many of you, when you go back to the place where you were raised, realize how few species and habitats remain?

MICHAEL SCOTT: I think the biggest challenge to protecting biological diversity in the United States today is to make sure we build bridges between the various constituencies, work together in a positive fashion, and realize that there might have to be compromises. In western states, excepting California, the opportunity still exists to make sound biological decisions that aren't driven solely by political considerations.

ANN HOOVER (USDA FOREST SERVICE): I work for the Forest Service, but I am also a graduate student at the State University of New York where I'm looking at ways of creating institutional corridors to preserve wildlife corridors. I've heard many of you refer to the need for increased cooperation between agencies and various groups, and I'm wondering if any of you can describe some mechanisms by which that cooperation might be had.

HAL SALWASSER: One of the things that the Forest Service has been using quite a bit recently is participating agreements and memoranda of understanding. For example, we have a memorandum of understanding with all forty-two state fish and wildlife agencies and formal arrangements with all the state foresters to coordinate our various plans and programs. We've expanded that with partnership agreements with The Nature Conservancy, for example, and a number of those agreements bear directly on biological diversity. The Rocky Mountain Elk Foundation, Trout Unlimited, and others share with us mutual goals. Between agencies, we have institutional arrangements. For example, adjoining land management units, like parks and forests, have arrangements that range from "gentlemen's agreements" up to some fairly formal congressionally mandated arrangements. The coordination that is going on in the Yellowstone area is being carried out by some interagency, interinstitutional agreements that didn't take an act of Congress to pull off.

FELICE PACE: To some people, cooperation means sitting down and talking. That's great, but that's not really cooperation. In my experience, the only time I see the major players in the timber industry, Forest Service, BLM, and the conservation community sitting down and talking is when somebody's interest is threatened and you have a club hanging over their head. That's unfortunate, but it's human nature to pursue our own interest until that interest is threatened. Sitting down and talking does not go far enough. There's too much emphasis on the process and not enough on the context.

BEN BROWN: At The Nature Conservancy, we've found that nothing gets done unless somebody is in charge. And, quite frequently, when you're dealing with an agency—particularly over a specific problem— the responsibilities of the people you're working with are so diverse that they can only spend a small part of their time on the problem. To counteract this, we've had some success with hiring temporary, full-time coordinators—people who coordinate the activities of the agencies in planning, budgeting, on-the-ground management, and so on. The cost is minimal when compared with the reward of a completed project.

There's another mechanism that sometimes works. Agencies have money. Ideally, they'd like to put someone on the problem full time, but they can't because they're dealing with a personnel ceiling or a hiring freeze. We found that interagency personnel agreements are often effective. The agency gives us the money, we hire somebody— usually for less salary than they would pay if they had to put this person to work full time.

Another system that has worked well is a protection planning committee, which may have another name now. These are basically inter-agency groups that meet anywhere from one to four times a year to talk about specific problems that are usually identified by a scorecard process. These problems often get solved very quickly—especially if a member of the committee has a direct line to the head of the agency that oversees the problem area.

KEITH HAY: There's no greater example of an institutional corridor than in your own state of New York. It's called the Hudson River Greenway and it encompasses three states, fourteen counties, 150 miles, and I don't know how many different agencies working cooperatively together.

II

Conservation Corridors: Countering Habitat Fragmentation

Introduction

BLAIR CSUTI

CONSERVATION BIOLOGY is a rapidly evolving discipline drawing on an ever widening number of biological, physical, and social sciences for concepts and applications. A decade ago, most emphasis was placed on population survival, based in large part on island biogeographic and genetic theory. (See, for example, Soulé and Wilcox 1980.) In the last several years, however, populations have increasingly been viewed in the context of communities, ecosystems, and now landscapes. While landscape ecology is a long-established discipline in Europe, the term "landscape" has appeared only recently in North American conservation literature (Naveh 1982). Since the terms ecosystem and community can be applied to areas ranging in size from the very small to the very large (Whittaker 1970; Norse 1990), "landscape" is the preferable term for describing large natural areas with conservation value. A landscape is defined as "a heterogeneous land area composed of a cluster of interacting ecosystems that is repeated in similar form throughout" (Forman and Godron 1986:11).

Placing conservation efforts in the larger spatial scale of landscapes has stimulated considerable discussion of corridors and landscape linkages. (See Noss 1983, 1987a; Noss and Harris 1986; Simberloff and Cox 1987.) The essays in this part of the book examine different temporal and spatial aspects of the definition and role of conservation corridors.

CONSERVATION TIMELINES

Scientists who study natural processes often have a different view of time than the administrators who set environmental policy. Geologists

81

place the age of the earth at 5 to 10 billion years. Anthropologists place the separation of the genus *Homo* from the great apes about 2 to 5 million years ago. Astronomers assure us we have another several billion years before we're deep-fried by an exploding sun. Politicians tell us they'll put a chicken in every pot just after the next election. Landscape dynamics—the random timing and location of disturbance events (like fires or floods) and a community's recovery from such events through succession—operate on a time scale of decades to centuries. The value of connectivity between landscapes should therefore be assessed from a long-term planning horizon: centuries rather than years.

Our species is going to make the planet earth its home for a long time to come. Faced with this reality, it makes sense to plan for long-term stability of the biosphere. At present, the loss of biodiversity through habitat fragmentation is accelerating rather than declining, both in temperate and tropical zones (Myers 1986). The endpoint of this downward spiral is biotic impoverishment and the accompanying breakdown of functioning ecosystems, which may trigger the collapse of civilization (Ehrlich and Ehrlich 1987). In Chapter 4, Hal Salwasser suggested that "we must often select options that only reduce the rate of loss." Given the consequences of continued erosion of our natural resource base, accepting loss, even at a reduced rate, is not a viable strategy for maintaining biodiversity. Instead, we can plan for stabilization of what's left of the world's biota while simultaneously meeting human resource needs. For some this means that expectations of natural resources must be brought into line with the realities of sustainable yields.

Since biological systems are inherently dynamic, the demise of local populations will continue (Soulé 1983) and the distribution of seral stages of vegetation types will shift over the landscape (through the centuries for such ecosystems as old-growth coniferous forests). Any comprehensive conservation strategy should move toward a steady state for natural systems that anticipates local disturbances while maintaining the overall proportion of system components. We must set our sights beyond individual populations and even species if we are to develop a strategy for the long-term maintenance of biodiversity.

From a temporal perspective, the contribution of corridors and landscape linkages to a conservation strategy must be assessed in evolutionary as well as ecological time. Frankel (1983) has pointed out that effective conservation measures can operate within shorter time scales—scales more in tune with human biology and culture. This does not diminish the importance of long-term maintenance of biodiversity. The distant benefits of maintaining evolutionary potential by retaining biodiversity are supplemented by immediate ecosystem support services supplied by healthy natural landscapes. Good long-term conservation planning is also good short-term economic planning.

CORRIDORS AND LANDSCAPE LINKAGES

Natural disturbances have always resulted in some degree of habitat fragmentation. In the past, disturbed patches were generally surrounded by a "sea" of later successional stages. Animals simply moved *around* islands of disturbance. Plants and plant communities were the matrix within which disturbance occurred. Recent human alterations in land cover have reversed this situation, however, and, in many areas, more or less permanent disturbances dominate the landscape.

There is general agreement that a "wildlife corridor" is a somewhat linear area of natural vegetation that connects larger areas of natural vegetation. In Chapter 6, Michael Soulé assigns a specific function to "conservation corridors"—the movement of animals. In Chapter 7, a practical application of conservation corridors, Felice Pace describes a series of landscape corridors, each "large enough to encompass large natural disturbances." In Chapter 8, Larry D. Harris and Kevin Atkins broaden their definition to include individual animal movements, gene flow, or the changing distribution of entire communities. Each of these three types of corridors meets Soulé's requirement of facilitating animal movement, but those described by Harris and Atkins and by Pace also accommodate slower movements of biodiversity elements (genes, communities, and processes). In fact, because Soulé's "movement" corridor is for transitional use only, it need not contain all the habitat elements required for long-term survival or reproduction of its migrants. Where possible, we should avoid divorcing landscape linkages for plants and small animals from movement corridors for mobile terrestrial vertebrates.

A landscape linkage differs from a movement corridor in that the complete range of community and ecosystem processes continues to operate within it through time, thus enabling plants and smaller animals to move between larger landscapes over a period of generations. Such linkages could provide an avenue for natural dispersal of communities in response to global climate change, as suggested by Harris and Atkins.

While the timing and magnitude of global warming are still debated, it is now ranked as one of the major threats to biodiversity (Peters 1988). While more mobile animals can seek out favorable habitats, plants and smaller animals can track changing habitats only over a period of generations. Landscape-scale linkages are the only possible avenue for such species to naturally follow the changing locations of favorable habitats. While artificial translocation of some species is possible, management alternatives are unlikely to move more than a fraction of the plant and animal diversity of a complete bioregion. Latitudinal and altitudinal landscape linkages remain the only practical remedy to the threat of global warming, and even they will prove wanting for species whose rate of dispersal is slower than the rate at which favorable habitat moves.

A corridor's function depends on its width, for edge effects (Janzen 1986) can penetrate some distance into corridors, depending on the habitat type. Edge effects typically can be measured 200 meters into a forest (Temple and Cary 1988) and sometimes as much as 600 meters from the edge (Wilcove et al. 1986). Using the 600-meter estimate, corridors narrower than 1.2 kilometers will contain no true interior habitat. Pace sets a minimum corridor width of 6.4 kilometers, wide enough to mitigate most edge effects. Soulé cautions, however, that very wide corridors can result in higher mortality of individuals in transit because random wandering lengthens travel time and increases exposure to predation and other threats. The discrepancy is rooted in different views of corridor function: If a species maintains a population within a corridor, some individuals will eventually die natural deaths. But the corridor can make a positive contribution to metapopulation dynamics if it is large enough to be a population source rather than a sink. That width depends on the species.

Pace makes the important point that corridors themselves are subject to natural disturbance events. To counter the threat that a corridor might be severed by a single natural disturbance, he suggests that corridors be wide enough to "encompass large natural disturbances." Not only do effective corridors link landscapes, they themselves have the characteristics of landscapes. As such, they go beyond the function of providing travel corridors for large animal species and can become important components of a reserve system. Since the magnitude and distribution of disturbance events are unpredictable, it is prudent to plan for multiple landscape linkages. Such a network of preserves connected by corridors (Noss 1987b) could approach prefragmentation levels of biological interchange within a regional landscape.

HOW THE BAD GUYS GET FROM HERE TO THERE

If corridors provide avenues for dispersal of indigenous species, they likewise facilitate the range expansion of nonnative species. Human-modified landscapes provide habitat for a number of opportunistic species, and they often reach these new habitats by way of natural corridors. The blue jay, for example, has crossed the Great Plains along the riparian corridors of the Platte River and can now be found in the urban and agricultural communities along the Rocky Mountain Front (Knopf, in press). Such additions to established communities can result in disruption of community structure and loss of diversity through competition or predation. The ubiquitous presence of species adapted to human-modified habitats at the expense of locally adapted forms diminishes regional diversity. To the extent that corridors facilitate such invasions, they play an adverse conservation role. This problem seems to be endemic to human-altered landscapes, since species rarely expand so quickly into well-established ecosystems under natural conditions.

Not only larger plants and animals but diseases, parasites, and pathogens can move from one natural area to another through corridors (Simberloff and Cox 1987). Prefragmentation landscapes displayed even higher levels of connectivity, suggesting that transport of parasites and pathogens was originally a local rather than a regional problem. Today there is an undeniable risk that parasites and disease agents can reach susceptible populations through corridors. While the loss of a local population is not usually devastating for a species, the value of linking unique populations to other landscapes may be outweighed by the risks of contamination.

Despite these potential disadvantages, corridors offer the only means of maintaining connectivity between habitat islands for mobile terrestrial animals (Noss 1987a). Such connectivity is essential for the operation of metapopulation dynamics. For most species and communities, linkage and the opportunity to function as part of a set of populations has a net positive effect.

EVOLUTIONARY EFFECTS

"Relaxation"—the gradual loss of species from a habitat island—is a well-documented phenomenon (Wilcove et al. 1986). Area-sensitive species, particularly the largest ones, are often the first to go. Arguing against movement corridors for smaller animals—those with large populations in relatively small areas—Soulé points out that they have a high probability of long-term persistence even without interlandscape dispersal. This is true in ecological time, but there are evolutionary consequences of isolation that argue for maintaining connectivity for plants and smaller animals.

The most conspicuous examples of adaptive radiation come from isolated oceanic islands such as the Galápagos and Hawaiian Islands (Bowman 1961; Carlquist 1970). In the absence of genetic exchange, selection to local conditions, combined with random changes in gene frequencies, leads toward the creation of reproductively isolated populations (Liu and Godt 1983). If the consequence of fragmentation is ultimately the creation of more species, would this counter the effects of anthropogenic extinctions? Unfortunately, the answer is no.

There is general agreement that biodiversity is a combination of community, species, and genetic diversity. Because of the difficulty of measuring genetic diversity, it is rarely addressed in conservation planning. We assume, however, that most species contain some degree of genetic variability and, moreover, that it is not uniformly distributed throughout the species' range. Isolated populations usually contain less than the full range of genetic variability present in the species. Certain genetic traits (alleles) are often geographically restricted to a subset of populations that comprise a species. If these isolates diverge to the point of reproductive isolation, intraspecific genetic variation is converted to interspecific genetic variation, each daughter species having less total variability than the parent species.

Another general point of agreement is that any single population has a certain probability of survival for a specified number of years. Given a long enough time, all populations will probably decline to zero. Species persist in spite of local population extinctions because of recolonization by nearby populations (Soulé 1983; Goodman 1987). When isolated species go extinct, they take with them a portion of the genetic variability of their parent species that cannot be replaced (McCauley 1991). If and when their habitat is recolonized by closely related species with a similar niche, the net trend in genetic variability for the larger evolutionary unit will be downward. Since extinctions are presently occurring much faster than useful mutations can accumulate in isolated populations, habitat fragmentation and isolation can only further erode global biodiversity.

FROM LANDSCAPE LINKAGES TO BIOSPHERE RESERVES

There is ample evidence that habitat fragmentation rends the genetic and demographic fabric of species, communities, and landscapes. Once, human disturbances were surrounded by wilderness; increasingly the wilderness is surrounded by development. Isolation is the ultimate fragmentation. Habitat "bridges" between habitat islands are the most direct remedy for the ills of isolation. Because of differing biological attributes, the value of corridors changes for each species using them. And because corridors are bounded by dissimilar habitat, their edges present physical and biological hazards to transients.

Relatively narrow corridors can facilitate the movement of larger animals. In the short term these are a necessary minimum to maintain many area-sensitive species. Since narrow corridors are subject to disruption by natural disturbances, more than one linkage between isolated natural areas is needed to maintain the benefits of connectivity through time. Broader corridors that display landscape-level processes, such as patch dynamics (Pickett and Thompson 1978), are required if plants and less mobile animals are to maintain contact with other populations of their species. The transit of these components of biodiversity through the landscape may take generations or, in the case of early successional species, even centuries.

A best-case scenario would have a network of large natural areas connected by landscape-scale linkages (Noss 1987b). The locations of natural areas that capture the greatest amount of a bioregion's diversity without redundancy can be determined through analyses of distribution maps of such indicators of biodiversity as vegetation, vertebrates, and butterflies (Pyle 1982; Scott et al. 1987, 1988). The transit of biodiversity elements within this network through landscape linkages would maintain ecological and evolutionary processes. Human activities would take place in intensively used lands within this network; there would be less invasive use of buffer areas around preserves and linkages (Noss and Harris 1986; Noss 1987b). Since natural disturbances will continue in such a system, the exact location of biodiversity elements (species, communities, seral stages) will change (Pickett and Thompson 1978), but there will be no net loss of species or ecosystems.

This ideal may be more or less attainable, depending on patterns of natural disturbance regimes, and may require management intervention (Baker 1989). If future changes in land use were planned to maintain the connectivity of wildlands, whole regions or even continents could act as functional biosphere reserves: a planned mosaic of exploited, seminatural, and protected areas that function both economically and biologically (Noss 1983; Noss and Harris 1986). Given our dependence on biodiversity services, the long-term social, economic, and cultural survival of our civilization would be well served by implementing such a system of linked biodiversity reserves.

REFERENCES

Baker, W. L. 1989. Landscape ecology and nature reserve design in the Boundary Waters Canoe Area, Minnesota. *Ecology* 70(1):23–35.

Bowman, R. I. 1961. Morphological differentiation and adaptation in the Galapagos finches. *Univ. Calif. Publ. Zool.* 58:1–302.

Carlquist, S. 1970. *Hawaii: a natural history.* Garden City, N.J.: Natural History Press.

Ehrlich, A. H., and P. R. Ehrlich. 1987. *Earth.* New York: Franklin Watts.

Ehrlich, P. R., and A. H. Ehrlich. 1981. *Extinction: the causes and consequences of the disappearance of species.* New York: Random House.

Forman, R.T.T., and M. Godron. 1986. *Landscape ecology.* New York: Wiley.

Frankel, O. H. 1983. The place of management in conservation. Pp. 1–14 in *Genetics and conservation: a reference manual for managing wild animal and plant populations* (C. M. Schonewald-Cox, S. M. Chambers, B. MacBryde, and W. L. Thomas, eds.). Menlo Park, Calif.: Benjamin/Cummings.

Goodman, D. 1987. How do any species persist?: lessons for conservation biology. *Conserv. Biol.* 1:59–62.

Janzen, D. 1986. The eternal external threat. Pp. 286–303 in *Conservation biology: the science of scarcity and diversity* (M. E. Soulé, ed.). Sunderland, Mass.: Sinauer.

Knopf, F. In press. Conserving diversity in cosmopolitan faunas. Unpublished MS.

Liu, E. H., and M. W. Godt. 1983. The differentiation of populations over short distances. Pp. 78–95 in *Genetics and conservation: a reference manual for managing wild animal and plant populations* (C. M. Schonewald-Cox, S. M. Chambers, B. MacBryde, and W. L. Thomas, eds.). Menlo Park, Calif.: Benjamin/Cummings.

McCauley, D. E. 1991. Genetic consequences of local population extinction and recolonization. *Trends in Ecology and Evolution*, vol. 6.

Myers, N. 1986. Tropical deforestation and a mega-extinction spasm. Pp. 394–409 in *Conservation biology: the science of scarcity and diversity* (M. E. Soulé, ed.). Sunderland, Mass.: Sinauer.

Naveh, Z. 1982. Landscape ecology as an emerging branch of human ecosystem science. Pp. 189–237 in *Advances in ecological research* (A. Macfadyen and E. D. Ford, eds.), vol. 12. London: Academic Press.

Norse, E. 1990. *Ancient forests of the Pacific Northwest.* Washington: Island Press.

Noss, R. F. 1983. A regional landscape approach to maintain biodiversity. *BioScience* 33:700–706.

————. 1987a. Corridors in real landscapes: a reply to Simberloff and Cox. *Conserv. Biol.* 1:159–164.

————. 1987b. Protecting natural areas in fragmented landscapes. *Nat. Areas J.* 7(1):2–13.

Noss, R. F., and L. D. Harris. 1986. Nodes, networks, and MUMs: preserving diversity at all scales. *Environ. Management* 10:299–309.

Peters, R. L. II. 1988. The effect of global climatic change on natural communities. Pp. 450–461 in *Biodiversity* (E. O. Wilson, ed.). Washington: National Academy Press.

Pickett, S. T. A., and J. N. Thompson. 1978. Patch dynamics and the design of nature reserves. *Biol. Conserv.* 13:27–37.

Pyle, R. M. 1982. Butterfly eco-geography and biological conservation in Washington. *Atala* 8:1–26.

Scott, J. M., B. Csuti, and K. Smith. 1990. Playing Noah while paying the devil. *Bull. Ecol. Soc. Am.* 71:156–159.

Scott, J. M., B. Csuti, J. D. Jacobi, and J. E. Estes. 1987. Species richness: a geographic approach to protecting future biological diversity. *BioScience* 37:782–788.

Scott, J. M., B. Csuti, K. Smith, J. E. Estes, and S. Caicco. 1988. Beyond endangered species: an integrated conservation strategy for the preservation of biological diversity. *Endangered Species Update* 5(10):43–48.

Simberloff, D., and J. Cox. 1987. Consequences and costs of conservation corridors. *Conserv. Biol.* 1:63–71.

Soulé, M. E. 1983. What do we really know about extinction? Pp. 111–124 in *Genetics and conservation: a reference for managing wild animal and plant populations* (C. M. Schonewald-Cox, S. M. Chambers, B. MacBryde, and W. L. Thomas, eds.). Menlo Park, Calif.: Benjamin/Cummings.

_____. 1985. What is conservation biology? *BioScience* 35:727–734.

Soulé, M. E., and B. A. Wilcox. 1980. *Conservation biology: an evolutionary-ecological perspective.* Sunderland, Mass.: Sinauer.

Temple, S. A., and J. R. Cary. 1988. Modeling dynamics of habitat-interior bird populations in fragmented landscapes. *Conserv. Biol.* 2:340–347.

Whittaker, R. H. 1970. *Communities and ecosystems.* London: Macmillan.

Wilcove, D. S., C. H. McLennan, and A. P. Dobson. 1986. Habitat fragmentation in the temperate zone. Pp. 237–256 in *Conservation biology: the science of scarcity and diversity* (M. E. Soulé, ed.). Sunderland, Mass.: Sinauer.

6 Theory and Strategy

Michael E. Soulé

CONSIDER THE FOLLOWING SCENARIO: Two mountainous areas are separated by lowlands that have been highly modified by road construction and agricultural conversion. Conservationists, believing that isolation of the biotas in the two areas is unnatural, press for a conservation corridor linking the two mountain ranges. Biologists, however, respond that the biological objectives of the conservation corridor need to be stated explicitly and claim that the desire for a corridor, while perhaps satisfying some subjective sense of landscape harmony, is too vague to serve as a basis for land use policy. Specificity is a necessity, for one cannot evaluate the success of such a project unless its objectives are clear and measurable.

Before describing the process of defining corridor objectives, however, it is best to define what I mean by "conservation corridor." There are many kinds of landscape corridors. (See Forman and Godron 1986; Harris and Gallagher 1989; Saunders and Hobbs 1991.) Some, like highway verges and utility rights-of-way, are incidental products of other land uses; others, like greenways, are primarily for human recreation; still others are simply elongated wildlife protection zones. The kind of corridor that I am discussing here—the conservation (or wildlife) corridor—has a very specific function: the movement of animals between two or more patches of habitat. (Corridors for plants deserve equal attention, but I leave this issue to the botanists.) These patches are here referred to as habitat islands, fragments, biodiversity islands, remnants, and nature reserves, more or less interchangeably.

More precisely, a conservation corridor is a linear landscape feature that facilitates the biologically effective transport of animals between

91

larger patches of habitat dedicated to conservation functions. Such corridors may facilitate several kinds of traffic, including frequent foraging movements, seasonal migrations, or the once-in-a-lifetime dispersal of juvenile animals. It is sometimes tempting to refer to any linear landscape element as a conservation corridor, but whether it actually functions as one depends on the corridor's capability.

Ultimately, the objective of such a conservation corridor is to increase the likelihood that a given species (or set of species) will persist in the habitat island and in the region. For many purposes, corridors may increase the odds of persistence by providing fresh individuals—immigrants—who can rescue a population from a crisis of too few individuals (Brown and Kodric-Brown 1977) or from inbreeding (Frankel and Soulé 1981). Immigrants can also recolonize an "empty" habitat patch if the target species is locally extinct. The important thing is to state the corridor's objective in terms of the viability of specific target species.

Some conservation corridors can provide essential ecosystem services to their associated isolates, whether or not the corridor serves demographic and genetic roles. A riparian corridor, for instance, can provide water, nutrients, and energy inputs to downstream fragments. In this chapter, however, the discussion is restricted to the demographic and genetic functions of corridors.

Most animals have very specific ecological requirements; these depend on their sex, age, condition, evolutionary history, and other factors. The habitat of a species (or habitats if it is migratory) must supply all of the resources and satisfy all of its life-history requirements. A corridor, in contrast, need not supply all the necessities of life. A corridor is a *transitional* habitat; it need only provide those ecological services and resources required when the individuals are moving between patches. For some species, the surrounding matrix of denatured habitat may be capable of providing some resources for migrating or dispersing individuals.

A corridor must be tailored to the needs of the species it is designed to serve, but it must not compromise the viability of other species. A corridor that functions well for deer, for example, may be harmful for salamanders. A poorly functioning corridor can do more harm than good because it can be a death trap for dispersing individuals—draining off healthy animals from a source area but failing to deliver them in sufficient quantity to the target patch. Such corridors can be called "sink corridors" because they are a net drain on the population and a hazard to the species. Designers must consider the possible impacts of a corridor on nontarget species as well as target species.

In summary, a hierarchy of questions must be addressed when considering the implementation of corridors for wildlife. The first is: What are the conservation objectives of the isolates? If one of the management objectives is to maintain viable populations of particular species, the second question must be: Which of these species may need a corridor to guarantee their persistence in the region, particularly in the isolates? The third question is: Can the corridor be designed so that it functions effectively without harmful side effects? Thus what started out as a simple landscape issue—connecting two mountain ranges—has turned into a complex problem of applied ecology.

SELECTING TARGET SPECIES

How does one select ecologically appropriate candidate species for corridors and avoid developing the right solution for the wrong problem? Since some species are obviously more in need of landscape linkages than others, one could try listing all the species in the region and, one by one, assess their respective need for corridors. Given that there are often hundreds or even thousands of macroscopic species in a region, this tedious approach is not very efficient. A backdoor way of selecting corridor species might be to choose them according to the kind of habitat that currently exists or could be restored in the potential corridor. This approach is flawed, however, because it substitutes convenience for an objective assessment of need. An established riparian corridor, for example, could have the wrong spatial orientation or lack the resources or microhabitats needed by a target species.

There are three general categories of need. First, some animals need corridors to facilitate periodic migrations to breeding or birthing sites. Second, some species may need to move between patches in order to forage or roost or to follow seasonally moving resources. Third, some populations must receive immigrants if they are to persist in isolated patches. Such a population is vulnerable because of the high probability of extinction associated with artificial isolation (Soulé 1987). An isolated population will not persist indefinitely (MacArthur and Wilson 1967). On average, the smaller the population, the less time it will persist (or the higher its chances of local extinction within any given interval); see Shaffer (1987) for a review.

Many factors are involved in the vulnerability of small, isolated populations. Chance events are the major villain. Chance operates at many levels—from catastrophes such as volcanic eruptions to less dramatic environmental fluctuations such as droughts and on down to the levels of genetics (inbreeding) and demography. By facilitating immigration, conservation corridors can dampen the impact of these chance events, significantly reducing the risk of local extinction due to isolation. In the rest of this section, I refer mostly to the needs of these isolation-sensitive populations.

Is there a shortcut to identifying animal populations of this latter type—those with the greatest risk of isolation-associated extinction? In other words, are there criteria that can be used to discriminate between populations at risk and those not so vulnerable in isolated patches of habitat? Such criteria do exist: abundance, variability in population size, and mobility.

Vulnerability. Table 6.1 lists some of the criteria that should be used when selecting target species for corridors. Risk of extinction, it must be emphasized, is directly related to rarity for demographic as well as genetic reasons. (See Terborgh and Winter 1980; Soulé 1987; Soulé et al. 1988; see Rabinowitz et al. 1986 for a discussion of the types of rarity.) Abundance, therefore, is a key criterion in the nomination of candidate species—the rarest animals should be at the top of the corridor candidate list. (The list of candidate species can be further refined by consulting experts in population viability. They will take into account other factors, such as variability in population size from year to year, migratory routes, and the distribution of resources among fragments.) Another indicator of vulnerability is the variability of population size (or growth rate) from year to year or generation to generation. High variability is associated with increased risk of extinction (Ehrlich 1983; Goodman 1987; Karr 1982).

Roughly speaking, large animals are rarer—many orders of magnitude rarer—than small animals. Among mammals, there is a fairly consistent relationship between body size and density, especially for carnivores: The bigger the animal, the less its density or the larger the home range (Harestad and Bunnell 1979; Eisenberg 1980; Calder 1984). Hence body size is a useful surrogate for abundance and can be used in a preliminary ranking of corridor candidate species. In most cases, candidates will be the largest native carnivores in the region, followed by the largest herbivores (see Brown 1971). One problem with this suggestion is that we often lack data on abundance of rare species. Usually, though, there is information on the population density of the animals in question—if not at the site, then at some place with similar habitat. Abundance can easily be estimated from density because density (D) and patch size (A) determine abundance (N): $N = DA$.

TABLE 6.1. Criteria for Selecting Target Species

Criteria for Candidacy	Contributing Factors
Abundance (N)	Area (A) of isolated patches and remnants Population density (D) Body size
Variability of population size	Body size, mortality, and fecundity rates
Mobility	Foraging needs (migratory) Breeding requirements (migratory)

How many candidate species will there be? It depends on the number of rare species. In most cases, common species need not be considered as corridor candidates because their numbers provide them with a margin of safety and buffer them from chance environmental events and random demographic and genetic events. Establishing corridors for perennially abundant animals is biologically superfluous. If the average fragment size is small, however, many of the species in the patches will be at risk. Patch size, therefore, is a major factor determining the number of target species as well as their identity. It is logical to assume that small habitat fragments will have lower population sizes than larger fragments. As illustrated in Figure 6.1, where average patch size is small, corridors must be capable of transporting species that are normally quite abundant, including insects, reptiles, amphibians, small mammals, and certain birds.

What if the patches to be connected are so small that the sum of their areas cannot support large animals? In such situations it might make sense to ignore large animals. For instance, it would be quixotic to plan a corridor that would accommodate grizzly bears in a system where the total area of the patches could not sustain even one pair of adults. But where many small patches add up to an area large enough to

FIGURE 6.1 A Guide to Selecting Target Species.

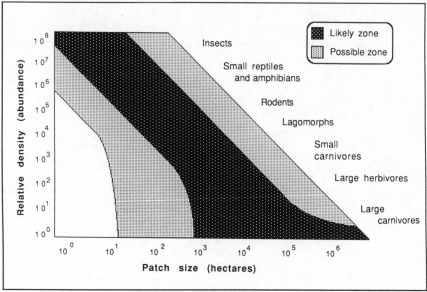

The selection of candidate species for corridors can be based on the likelihood of their extinction within the patch—which is a function of their abundances and the size of the patch. Abundant (dense) species are likely candidates only when reserves are small. Large animal species often require corridors over a broader range of reserve sizes. The dark shading indicates the zones of abundance and patch sizes for which a species is likely to require a corridor. The lightly shaded areas represent zones where corridors may be necessary or helpful. This figure assumes that a minimum viable population for most animals is in the range 10^3 to 10^4 (Belovsky 1987).

support some large animals, a corridor might be justified. By connecting such small patches with a network of corridors, it might be possible to maintain a viable population of a large animal, at least for a few years or decades.

Extreme caution should be exercised in excluding large animals. Large animals are often "keystone" species, strongly interacting with many other members of the community despite their relative rarity. Such keystone species may be critical for the maintenance of habitat and species diversity (Mills and Soulé 1991). It is often ecologically prudent, therefore, to provide access to large, rare animals in systems where no single patch could support a permanent resident or population. It may even make good sense to maintain such large animals in a system when expensive management interventions will be necessary to sustain them—for the disappearance of large animals often leads to the decline and extirpation of many smaller animals (Soulé et al. 1988; Terborgh 1988). The keystone role of large animals, particularly predators, is indicated in Figure 6.1 by the wide range of abundances over which corridors are shown to be desirable for them.

Isolation. The last phenomenon to be considered is isolation. Isolation is usually measured as the distance (or the square of the distance) between patches. It can vary from a meter or so to many kilometers. How does distance between patches—which is the same as minimum corridor length—affect the choice of species that might use the corridor? First, the longer the corridor, the faster the animals must move (mortality rates being equal). Stated slightly differently: The higher the mortality rate in the corridor, the faster the animals must move, or the shorter the corridor should be, everything else being the same. In other words, two rates, mortality and speed, determine the maximum length of a corridor.

Speed depends on the style of locomotion. Burrowing animals take hours or days to move 1 meter, whereas a dispersing bird might cover 100 kilometers or more in the same interval. The range in speed among almost all the species in a region spans as many orders of magnitude as the ranges in body size and density—about nine. For our purposes, we can simply observe that a 10-kilometer corridor is too long for frogs, especially if in the presence of hungry raccoons.

Summary and Applications. The application of these principles to the selection of candidate species for corridor use can be expressed as follows: The major criterion in the identification of candidate species is their vulnerability to extinction in the patches to be linked by corridors. And a principal criterion of vulnerability is abundance based on average population density. Total abundance is usually proportional to the size of the patches. In many cases body size and abundance are inversely correlated, at least within a set of related species. As shown in Figure 6.1, the larger the patches, the greater the average size of the animals that will be viable within them and the fewer the species that need be considered for corridor use.

The main point is that the *function* of a corridor must be stated explicitly; otherwise the corridor may do more harm than good (Soulé and Simberloff 1986). Returning to the two mountain ranges cited earlier, say that all parties agree that most of the animal populations in the mountains, with one or two exceptions, have viable populations or, like the raptors, have the capacity to cross the lowland barrier with ease. It follows, then, that it is unnecessary to design a corridor that would facilitate the movement of fish, frogs, rodents, small carnivores, migratory birds, and eagles. A lowland corridor designed to facilitate their conveyance would be a great waste of resources. One of the exceptions is the mountain lion or panther (*Felis concolor*), typically the least abundant top carnivore, whose population size puts it at extreme risk. Another exception might be a tortoise. Although tortoise density is often high, this was not the case in our mountains—in part because of high mortality from automobiles, in part because egg-laying sites are localized and rare.

DESIGNING THE CORRIDOR

Once the target species have been selected, the stage is set for the second step of corridor analysis: the design. (Only an outline of the principles of corridor capability can be attempted here; see also Harris and Eisenberg 1989 and Chapter 8 in this volume.) Whether the purpose of a corridor is to transport mountain lions or tortoises, the analysis of corridor capability must be examined through the lens of the species in its environment. That is, the analysis must be "autecological." Autecology, the study of the ecological interactions of a single species, overlaps greatly with the criteria for capability presented in Table 6.2.

Much of autecology concerns the factors that determine the local abundance of a species. In the previous section on candidate species, we discussed differences in abundance *between* species. Abundance is also relevant to corridor design and capability, but here we are concerned with the relative numbers of different groups *within* a species or how individuals are distributed between the two sexes and between age and size classes in both space and time. In many species, for example, juvenile males are more likely than other classes of individuals to make long-distance movements. This means that corridor designers may have to consider the factors that control the proportions of individuals that occur in the different sex and age classes of a population. These proportions in turn depend on an understanding of demographic vital

TABLE 6.2. Criteria for Designing the Corridor's Capability

Criteria for Capability	Contributing Factors
Species or population	Movement type and magnitude
	Demographic vital rates (birth and death)
	Movement rate (flight, cursorial, fossorial)
Environment	Habitat quality (abiotic and biotic factors)
	Barriers to movement
	Effects of edges on mortality
	Interactions between organisms (intraspecific and interspecific)

rates, including age-specific birth and death rates. Corridor design, therefore, must take into account the life history and demography of target species.

Next to be considered are the manifold intrinsic and extrinsic (environmental) factors that influence movement. A corridor will be avoided by a species if it is the wrong habitat, for example, or if there are barriers, physical or psychological, to movement. Other important considerations are the behavioral factors that affect movement including curiosity, navigational ability, learning capacity, the magnitude of movements, and the timing and periodicity of movements. Some of these factors are affected by intraspecific and interspecific interactions including territoriality, competition, and predation. A corridor's capability clearly depends on interactions between a species and the environmental conditions. In other words, most ecological factors can affect death rates within corridors and therefore the corridor's capability. The death rate of species in a corridor is the major criterion in determining its effectiveness. If death rates are high, it is unlikely that individuals will complete the journey.

Death rates will not be uniform. For many species, death rates will vary between the edge of the corridor and the interior. For example, a species that is vulnerable to predators may experience much higher mortality rates along edges because many predators (ravens, jays, raccoons, house cats, snakes, foxes) prefer to forage where their movement and visibility are least obstructed. The ratio of edge habitat to interior habitat is one aspect of the theory and topology of corridors that we (Soulé and Gilpin 1991) have begun to examine using a computer simulation. Our conclusions, in brief, are:

- Optimum corridor width depends on the strength of the edge effect: The higher the relative mortality rate of the edge, the wider the corridor must be. But very wide corridors are suboptimal because animals spend time wandering around inside the corridor, all the time suffering a certain mortality rate. Of course, the lower the mortality rate, the wider the corridor can be.
- Any departure from linearity may be deleterious; a straight corridor is superior because animals spend less time in edge habitat. A corridor with barriers, doglegs, cul-de-sacs, and the like has the character of a maze, even for relatively "intelligent" animals.
- Corridors are most effective with straight sides and a constant width. Funnel-shaped (gradually narrowing) or, in reverse, horn-shaped (gradually widening) corridors are less effective than those with straight sides and a constant width.

Since the theory of conservation corridors is a new discipline, many changes and controversies can be expected in these early years. We should not, however, lose sight of the main purpose of corridor capability analysis: Survival within the corridor must be high enough to justify its expense.

Consider, again, the scenario of the two mountain ranges and the conservation corridor that might link them. Biologists have now informed the planners that they can ignore the needs of toads, wallabies, and ravens because these animals are not appropriate target species. Moreover, the designers can safely ignore most of the resource needs of the mountain lions because their passage through the corridor will be swift and it is the core habitats in the mountain ranges, not the corridor, that provide food and shelter. Because mountain lions can cover a lot of ground quickly, and because they are habitat generalists and have no enemies except other mountain lions and humans, a lion corridor is relatively simple and cheap to develop. In many cases, the corridor may be merely a highway underpass, though in this scenario the development of the corridor would probably require the restoration and fencing of the approaches to the underpass in order to funnel the lions to the crossing point. For other kinds of species, however, habitat considerations will be more important—particularly if the species is smaller, less mobile, and generally more vulnerable to predation. Even small barriers or gaps in cover could be major hazards for forest-interior, grassland, or chaparral-type animals. A corridor that is capable and efficient for *both* mountain lions and tortoises will have to be much more carefully designed than one for lions alone.

Little will be said here about evaluation. The main aspect is the monitoring of the movement and the mortality of target species. It should be borne in mind that even an occasional successful disperser can benefit a genetically depauperate recipient population. But there are often demographic and economic costs associated with dispersal via a corridor. It may be cheaper and more effective, in some cases, to translocate animals artificially, though the attendant commitment to a long-term program may be more costly and the success rate too poor. Another problem is excessive mortality in the corridor—a low success rate may endanger the source population. For large, slowly reproducing animals, however, a well-designed corridor may greatly reduce the road kill mortality and more than justify the expense of a highway underpass or overpass.

WILL CORRIDORS WORK?

High rates of species extinction, local and general, can be compared to an epidemic. The cause of this epidemic is the fragmentation and destruction of wildlands and waters by denaturing activities such as logging, water projects, mining, road construction, agricultural conversion, urbanization, and similar human projects. (See Wilcove et al. 1986 for a review of fragmentation.) Fragmentation leads to the isolation and imprisonment of populations, and these are the harbingers of extinction. Conservation corridors can help to redress these problems, but only if they are planned, executed, and monitored with care and in light of research. The design of capable corridors is a team task. Biologists must join with engineers, planners, conservationists, and developers to produce a system of corridors that protects rather than destroys complex biological systems.

When fragmentation is occurring, planners and conservationists should initiate a crash program to determine the relative vulnerability of the local species and start planning for a system of corridors that will prevent local extinctions. Such a program should be strategic and clearly distinguish between the three phases of corridor development: target species selection, design and capability analysis, and evaluation. Target species selection is based on species-specific needs (such as routes for breeding migrations) as well as on abundance as shown in Table 6.1. It is essential, also, to be explicit about the objective. In practice, a corridor should be designed for a particular species or a well-defined group of species and should take into account the management objectives of the fragments to be connected.

The need for corridors can be based on guidelines such as those offered in Figure 6.1, but guidelines cannot replace common sense and expert knowledge. Guidelines and rules of thumb should stimulate thought, not replace it. For example, Figure 6.1 suggests that abundant animals such as insects or frogs will not require corridors if the habitat patches are large. But as pointed out earlier, many animals are migratory and use different habitats in different seasons. Corridors may be necessary to prevent the disappearance of some common species, even in relatively large patches.

In general, vulnerability to extinction within isolated patches is inversely related to their size. Hence if the patches themselves are very large, the corridor links between them need not be capable of transporting most small-bodied, abundant species. The converse is not necessarily true, however. Many small patches can often sustain larger animals, assuming that movement among the patches is possible. Coyotes (*Canis latrans*), for example, are adept at hop-scotching between isolated patches of chaparral in southern California, and the presence of keystone species such as coyotes can help to maintain local biological diversity and integrity (Soulé et al. 1988).

Corridor design is based on autecological knowledge, although the demands that animals make on corridor habitat will usually be less than they make on breeding or core habitat. The design of corridors that link small remnants is more difficult than that for large isolates. Corridors that link small remnants must safely carry small, relatively abundant animals (such as frogs and tortoises) as well as large, rarer ones (turkeys, badgers, coyotes). This is because large fluctuations in numbers are common in small animals, and local extinctions can occur (Ehrlich 1983) even when the long-term average size of a population is large. In such cases, corridors facilitate natural recolonization. Alternatives to corridors, however, should be examined at every stage. It should also be borne in mind that establishing corridors before fragmentation is complete is much less expensive than corridor therapy—retrofitting the landscape with corridors *after* denaturation has eliminated the natural habitat between remnants or reserves.

We are currently engaged in an immense, unplanned experiment that is testing the vulnerability of the biosphere to massive disruption. The experiment is crude but complex. How effective will corridors be in mitigating habitat or landscape fragmentation? No one knows yet. Much more work, theoretical and empirical, needs to be done.

ACKNOWLEDGMENTS

I wish to thank Merry Camhi for very helpful review of an earlier draft. William A. Calder III, Paul R. Ehrlich, and John Harte made many useful comments and suggestions.

REFERENCES

Belovsky, G. E. 1987. Extinction models and mammalian persistence. Pp. 35–58 in M. E. Soulé (ed.), *Viable populations for conservation*. Cambridge: Cambridge University Press.

Brown, J. H. 1971. Mammals on mountaintops: nonequilibrium insular biogeography. *Am. Naturalist* 105:467–478.

Brown, J. H., and A. Kodric-Brown. 1977. Turnover rates in insular biogeography: effect of immigration on extinction. *Ecology* 58:445–449.

Calder, W. A. 1984. *Size, function, and life history*. Cambridge: Harvard University Press.

Ehrlich, P. R. 1983. Genetics and extinction of butterfly populations. Pp. 152–163 in C. M. Schonewald-Cox et al. (eds.), *Genetics and conservation*. Menlo Park, Calif.: Benjamin/Cummings.

Eisenberg, J. F. 1980. The density and biomass of tropical mammals. Pp. 35–55 in M. E. Soulé and B. A. Wilcox (eds.), *Conservation biology: an evolutionary-ecological perspective*. Sunderland, Mass.: Sinauer.

Forman, R. T. T., and M. Godron. 1986. *Landscape ecology*. New York: Wiley.

Frankel, O. H., and M. E. Soulé. 1981. *Conservation and evolution*. Cambridge: Cambridge University Press.

Goodman, D. 1987. The demography of chance extinction. Pp. 11–34 in M. E. Soulé (ed.), *Viable populations for conservation*. Cambridge: Cambridge University Press.

Harestad, A. S., and F. L. Bunnell. 1979. Home range and body weight—a reevaluation. *Ecology* 60:389–402.

Harris, L. D., and J. Eisenberg. 1989. Enhanced linkages: necessary steps for success in conservation of faunal diversity. Pp. 166–181 in D. Western and

M. C. Pearl (eds.), *Conservation for the 21st century.* Oxford: Oxford University Press.

Harris, L. D., and P. B. Gallagher. 1989. New initiatives for wildlife conservation: the need for movement corridors. Pp. 11–34 in G. Mackintosh (ed.), *Preserving communities and corridors.* Washington: Defenders of Wildlife.

Karr, J. R. 1982. Population variability and extinction in the avifauna of a tropical land bridge island. *Ecology* 63:1975–1978.

MacArthur, R. H., and E. O. Wilson. 1967. *The theory of island biogeography.* Princeton: Princeton University Press.

Mills, L. S., and M. E. Soulé. 1991. The history and current status of the keystone species concept. Unpublished MS.

Rabinowitz, D., S. Cairns, and T. Dillon. 1986. Seven forms of rarity and their frequency in the flora of the British Isles. Pp. 182–204 in M. E. Soulé (ed.), *Conservation biology: science of scarcity and diversity.* Sunderland, Mass.: Sinauer.

Saunders, D. A., and R. J. Hobbs (eds.). 1991. *The role of corridors in nature conservation.* Sydney: Surrey Beatty & Sons.

Shaffer, M. 1987. Minimum viable populations: coping with uncertainty. Pp. 69–86 in M. E. Soulé (ed.), *Viable populations for conservation.* Cambridge: Cambridge University Press.

Soulé, M. E. (ed.). 1987. *Viable populations for conservation.* Cambridge: Cambridge University Press.

————. 1991. Land use planning for the maintenance of wildlife in a fragmenting urban landscape. *J. Am. Planning Assoc.* 57(3):312–322.

Soulé, M. E., and M. E. Gilpin. 1991. The theory of wildlife corridor capability. In D. A. Saunders and R. J. Hobbs (eds.), *The role of corridors in nature conservation.* Sydney: Surrey Beatty & Sons.

Soulé, M. E., and D. Simberloff. 1986. What do genetics and ecology tell us about the design of nature reserves? *Biol. Conserv.* 35:19–40.

Soulé, M. E., D. T. Bolger, A. C. Alberts, R. Sauvajot, J. Wright, M. Sorice, and S. Hill. 1988. Reconstructed dynamics of rapid extinctions of chaparral-requiring birds in urban habitat islands. *Conserv. Biol.* 2:75–92.

Terborgh, J. 1988. The big things that run the world—a sequel to E. O. Wilson. *Conserv. Biol.* 2:402–403.

Terborgh, J., and B. G. Winter. 1980. Some causes of extinction. Pp. 119–131 in M. E. Soulé and B. A. Wilcox (eds.), *Conservation biology: an ecological-evolutionary perspective.* Sunderland, Mass.: Sinauer.

Wilcove, D. S., C. H. McLellan, and A. P. Dobson. 1986. Habitat fragmentation in the temperate zone. Pp. 237–256 in M. E. Soulé (ed.), *Conservation biology: science of scarcity and diversity.* Sunderland, Mass.: Sinauer.

7 The Klamath Corridors: Preserving Biodiversity in the Klamath National Forest

Felice Pace

THIS CHAPTER INVESTIGATES the biological impacts of forest fragmentation resulting from logging and road building in the Klamath Mountain Province of southern Oregon and northern California. To maintain biodiversity in this region, I present a proposal to designate certain drainages and ridge systems in the Klamath National Forest as biological corridors linking designated wilderness areas. The proposal is designed as a strategy to ensure that the impact of human activity will not change current species, habitat, and genetic diversity within the Klamath National Forest over the long term. While the concept applies approaches developed by Larry Harris and Reed Noss, the proposal itself is the creation of the Klamath Forest Alliance, a coalition of grass-roots environmental and community organizations located in and around the Klamath Forest. At this time the proposal has not been implemented nor has it been endorsed by any individual or agency other than the Klamath Forest Alliance.

THE KLAMATH MOUNTAIN PROVINCE

The Klamath National Forest is located at the heart of what is often referred to as the Klamath Mountain Province. Spanning the California/Oregon border, this area of steeply folded, granite-cored mountains is strategically situated at the junction of the Northwest (or

Cascadian), Californian, and Great Basin bioregions. Within the province the maximum elevation is slightly over 9,000 feet, upland valleys occur in the 1,500–3,000-foot range, and major river canyons wind through the mountains to the Pacific shore.

Biologically speaking, plant diversity is the area's hallmark. A long list of plant species are endemic to the Klamath Mountains. Noteworthy in this regard are the herbaceous plants that grow only on serpentine soils of the Kalmiopsis and Siskiyou mountains. It is with respect to tree diversity, however, that the area's greatest biological significance can be found. Within 1 square mile of the Sugar Creek drainage in the Klamath National Forest, for example, seventeen species of coniferous trees are found (Sawyer et al. 1970). Nowhere on earth is there a greater diversity of conifers. The Klamath Forest Alliance believes that the unique plant (and especially conifer) diversity of the Klamath Mountains qualifies the area for international recognition and protection as a World Heritage Biosphere Preserve.

Biologists and naturalists have identified the factors contributing to the outstanding plant diversity of the Klamath Mountains. The province was dry land when most of California was still under the sea. Species dominant during older geological periods—species that now occur at no other sites in California—can be found in the Klamath Mountains. Currently the region's climate is transitional between the wetter clime characteristic of the Northwest and the Mediterranean clime characteristic of California. Consequently, many northwestern plants find the southernmost extent of their range in the Klamath Mountains. Likewise, the northern range limit for several Californian plants occurs here.

Tree ring and other data suggest a historic alternation of climatic cycles. During warm and dry cycles, southern and eastern plants have extended their ranges to the north and west. Colder and wetter times have allowed northern plants, especially forest trees, to extend their range southward. Several large river systems cut through the Klamath Mountains from east to west providing avenues for the dispersal of species. Of the four major river systems—the Trinity, Klamath, Rogue, and Umpqua—the Klamath is probably the most significant as an avenue of plant dispersal. Like the Columbia to the north, the Klamath cuts through the Cascade Range providing a low-elevation route for Great Basin plants to migrate westward during periods of dry weather.

Because the high relief and multidirectional ridge systems have produced a variety of microclimates, remnant plant populations have tended to persist in the Klamath Mountains when they were extirpated from surrounding regions. Indeed, the unusual confluence of factors favoring plant diversity in the Klamath Mountain Province has resulted in a large number of unique plant associations. Only in recent years have ecologists begun to identify and classify these associations, and the process is far from complete (Atzet and Wheeler 1984).

Although the Klamath Mountains are not as well known for animal diversity, they do provide suitable habitat for many species of concern, including the Pacific fisher, wolverine, pine marten, and northern spotted owl. From a wildlife perspective, the region is important more for its location as a link between provinces than for endemism. Patterns of development along the West Coast have tended to isolate forested regions. For example, agricultural, urban, and highway development in the interior valleys of Oregon has created barriers to migration of animals associated with forest environments between the Coast Range and the Cascades. A similar barrier has developed between the southern Oregon Cascades and the northern Sierra Nevada in California. Located at the geographical crossroads of West Coast forest regions, the Klamath Mountain Province must play a key role in any strategy to maintain and restore biological connectivity between natural forest reserves in the western coastal portion of the United States.

Most watersheds in the Klamath Province support commercially important runs of anadromous salmonids. These fisheries have been in decline ever since statistics were first recorded in the 1940s. Of special concern are remnant populations of spring chinook salmon and summer steelhead. Both are now found in only a handful of streams and at population levels that may not be viable over the long term. Anadromous salmonid populations are increasingly dominated by hatchery-bred fish. Some biologists have suggested that the genetic diversity contained in wild stocks of anadromous salmonids will be increasingly important as reliance on hatchery fish continues to grow (National Council on Gene Resources 1982).

FOREST FRAGMENTATION

Although logging began in the Klamath Province with the coming of white settlers, impacts on the forest ecosystem were minor prior to World War II. The development of industrial forestry after that war, fueled by the demand for new housing, has resulted in major alterations to the landscape. Initially selection was the dominant logging practice, but by the mid-1960s the USDA Forest Service and Bureau of Land Management, which administer the majority of the province's forests, had adopted clearcutting as the preferred timber extraction technique and tree plantations as the method for reforestation. For technical and economic reasons, high-density road systems have been constructed to reach timber in steep, mountainous terrain. An active program to suppress fires, whatever the cause, has also been in continuous operation since the early part of the century.

During those early decades when selection and other partial tree-removal systems were the dominant logging methods, federal agencies charged with forest management relied on natural seeding from residual trees on the site to reforest logged lands. Some structural diversity was retained, although the large trees were removed from logged sites. With the dominance of clearcutting, planting of nursery-raised seedlings has become the typical reforestation practice. The result is a forest of even-aged trees with little structural diversity. Artificial reforestation has been problematic throughout the Klamath Province, particularly when sites with shallow soils and southwestern exposures are clearcut. Some logged sites have been replanted three or four times without establishment of a forest stand (Perry et al. 1989). On some sites forest has been replaced by brush or grass as the dominant vegetation type.

In the Klamath Province fire plays a major role in the life history of native forest ecosystems. Natural forest fires typically burn in a mosaic pattern. For the most part they are underburns—understory vegetation is reduced and the stand is thinned as weaker and diseased trees are killed. Forest openings are created where the fires flare up (because of fuel, wind, and other factors) to kill patches and "stringers" of forest. In this way fire helps create a diversity of successional stages. Some foresters have theorized that clearcut forest management can act in the same way, creating a mosaic of stands of different ages. But comparison of natural and managed portions of the forest reveals basic differences in the size of openings and pattern of forest stages. More important, before human intervention the native forests of the Klamath Province were dominated by the older age classes whereas human management has resulted in the dominance of earlier successional stages.

Logging and fire suppression create unnaturally high levels of dead woody fuels in the forest. When a fire escapes immediate suppression in such an area, it often "blows up" into a firestorm that can kill all vegetation over a large area. Catastrophic fires have become more common and larger as the forest area under clearcut and plantation management has increased. A single fire complex in the Klamath Province can now cover more than 60,000 acres. Salvage logging and replanting after such fires have created vast plantations of young trees all approximately the same age and height. These plantations are highly susceptible to catastrophic fire.

The combination of extensive clearcut/plantation management, type conversions, and large catastrophic fires has created a forest landscape in the Klamath Mountain Province distinctly different from native conditions. In 1989 the Klamath Forest Alliance in cooperation with the Marble Mountain and National Audubon societies mapped the coniferous forests and tree plantations of the entire Klamath National Forest. The project utilized timber strata data compiled by the Forest Service from aerial photographs and representing ground conditions in December 1988. These maps demonstrate the extent to which forest fragmentation has progressed in the Klamath Forest. For the most part, wilderness areas appear as islands of natural habitat in a sea of clearcuts, young tree plantations, and isolated stands of older trees. Research referenced elsewhere in this book has amply demonstrated the limitations of small, isolated forest fragments as habitat for interior-forest species. The maps also reveal a few areas where the natural habitat connectivity of the forest has not been compromised. These watersheds and ridge systems have not yet been reached by the network of logging roads that covers most of the forest to facilitate the extraction of logs. They are the last remaining source of connectivity between reserved lands.

The spotted owl is perhaps the best-known species native to the Klamath Province whose fate is tied to forest fragmentation. But there are many other animal species still present in the province that find optimal conditions in forests dominated by old-growth trees. The Pacific fisher and other furbearers are of particular concern because of their solitary habits and extensive home ranges. Although such animals are considered "sensitive" by state and federal wildlife agencies and have been identified as "management indicator species" by the Forest Service, data on their status in the province is limited. Nevertheless, there are indications that conversion from native to managed forests in the Klamath Province and elsewhere is having a significant impact on population and other factors that bear directly on the long-term survival of these animals (Calif. Dept. of Fish & Game 1987).

PRESERVING BIODIVERSITY

In the absence of substantive requirements on private lands, responsibility for conserving biodiversity has fallen to agencies that manage public lands. In the Klamath Mountain Province over half the land is in public ownership, the majority administered by the USDA Forest Service as part of the National Forest System. Within these national forest lands are designated wilderness reserves ranging in size from the 500,000-acre Trinity Wilderness to the 4,000-acre Red Buttes Wilderness. Until recently it has been assumed that these wilderness areas would provide sufficient habitat to ensure maintenance of viable wildlife populations. Smaller tracts have also been reserved for federally listed threatened and endangered species under provisions of the Endangered Species Act and as research natural areas.

Reliance on wilderness areas, national parks, and other reserved habitat islands to preserve biodiversity has many problems. Newmark (1987) has noted the disappearance of mammalian species from some of our largest reserves. He concludes that most parks are too small to support the assemblage of mammals present when they were created. Most national parks and wilderness areas have been created with scenic and recreational concerns paramount. Wilderness areas recommended by the Forest Service have typically been designed to exclude commercially valuable forested areas. In the West these reserves occur for the most part at higher elevations, the "rock and ice" wildernesses. In the Klamath Mountain Province the subalpine and true fir forest types are fairly well represented in wilderness areas because these types predominate at elevations above 5,000 feet. The mixed conifer, Douglas fir, ponderosa pine, and mixed evergreen forest types, however, are not well represented in these reserved areas.

The National Forest Management Act (NFMA) specifically mandates preservation of current biodiversity on national forest lands. Managers are also charged by NFMA to ensure that viable populations of all naturally occurring wildlife are maintained. These provisions are in the process of being implemented, primarily through regional guides and land management plans (LMPs) for each national forest. Elsewhere in this volume Hal Salwasser describes the USDA Forest Service's policy on maintaining biodiversity in our national forests. In general, the approach continues to be to disperse biodiversity concerns across a managed landscape, minimizing reserved and special-use land.

The Forest Service's "New Perspectives" program will attempt to combat the impact of forest fragmentation by retaining structural diversity and other key elements of native ecosystems in managed timber stands. It is an encouraging development and deserves support. There are indications, however, that the Forest Service plans to use this program as justification for building roads and conducting logging in the last remaining watersheds and ridge systems that provide natural, unfragmented linkages between larger blocks of reserved lands. These proposals should be deferred until it can be demonstrated (through implementation and monitoring in areas where timber management has already taken place) that "New Perspectives" forestry truly maintains biodiversity.

The Forest Service has recognized that species dependent on native and old-growth forests require special management programs. In West Coast national forests, the biodiversity and viable population provisions of NFMA have been approached by identifying management indicator species (MISs) that stand for complexes of species dependent on particular ecosystems. In the case of Northwest and northern California native forest ecosystems, the northern spotted owl was selected as the indicator species. Forest Service biologists and managers have identified a system of habitat islands, ranging in size from 1,000 to 2,400 acres, intended to ensure maintenance of viable populations of spotted owls and all other species for which the spotted owl is considered the indicator. Creating smaller islands between the larger reserves is a strategy that attempts to combat the effects of forest fragmentation with a system of fragmented habitat islands.

In April 1990 an interagency group of eminent biologists charged by Congress to study the spotted owl's status called the Forest Service's habitat island plan "flawed" and recommended much larger habitat islands utilizing wherever possible lands already in reserves (Thomas et al. 1990). While the larger habitat islands are an improvement over the present small-island strategy, it remains a system that does not ensure connectivity of native forest habitat.

Controversy surrounding study of the spotted owl for listing as threatened or endangered under provisions of the Endangered Species Act has obscured the owl's assigned role as an indicator species. This role has itself been questioned by many biologists, most notably by the Interagency Spotted Owl Study Group. The group's report emphasized that the strategy proposed for the spotted owl would not necessarily ensure survival of all wildlife species that depend on native forest habitat. Many biologists believe the spotted owl is not a good indicator for the Pacific fisher and other terrestrial forest animals with extensive home ranges.

THE KLAMATH CORRIDORS PROPOSAL

The Klamath Corridors Proposal is based on the logical deduction that the most effective way to combat fragmentation is to retain or reestablish natural connectivity. The proposal makes maximum use of already reserved lands. Although they are not well represented at lower elevations, these lands are the largest blocks of unfragmented habitat remaining. The proposal also incorporates the history of human disturbance. The best landscape linkages available today are those few watersheds and ridge systems that remain substantially free of roads and clearcuts. While major rivers once provided the primary avenues for plant and animal dispersal, today human development centers on these rivers. Consequently, it is the more remote watersheds and ridge systems that now offer the best opportunities for maintaining, enhancing, or reestablishing the connectivity of native forest habitat.

In June 1989 the Klamath Forest Alliance, Klamath National Forest, and California Department of Fish and Game cosponsored a workshop to investigate strategies for preserving biodiversity in the Klamath Forest. Reed Noss and John Lehmkuhl, a USDA Forest Service research biologist, were featured speakers. Forest Service biologists and line officers from the Klamath National Forest, California Department of Fish and Game biologists, and members of environmental organizations participated. The three-day workshop included presentation of landscape approaches to maintaining biodiversity in fragmented landscapes and consideration of how such approaches could be used in the Klamath National Forest. Maps and aerial photographs were studied to determine the most appropriate natural links between reserved lands. A day was spent in the field looking at two of the drainages under study as corridors between wilderness areas.

Three types of corridors were discussed at the workshop: 50 to 200-foot-wide riparian corridors utilizing existing Streamside Management Zones (SMZs); one-quarter to 1-mile-wide riparian and ridge corridors to provide travel routes for dispersing animals; and large corridors (full drainages and ridge systems) to provide a continuum of functional native habitat connecting the large wilderness reserves. Although a draft prescription for the larger landscape corridors was developed at the workshop, there was no consensus on which type of corridor to recommend for use in the Klamath National Forest. Moreover, there were differing opinions on the efficacy of human "management activity" (fire suppression, vegetation manipulation, and so forth) within corridors. In general, the managers favored access for management while the conservationists favored no entry. (In a peer review of the workshop report, some questioned the effectiveness of corridors, particularly the narrower travel corridors.)

Using information from the June 1989 workshop and the conifer maps produced from Forest Service data, the Klamath Forest Alliance developed the Klamath Corridors Proposal. This proposal uses entire watersheds (and in some cases assemblages of watersheds) and ridge systems to link lands currently reserved as wilderness in the Klamath and adjacent national forests. Figure 7.1 is a map of the Klamath National Forest (except Goosenest District) showing the wilderness areas and the proposed corridors. The corridors are designed to be large enough to encompass large natural disturbances—for example, forest fires. When ridge systems were utilized, we aimed for a width of at least 4 miles adjusted to conform with topographic and vegetation boundary features.

Some critics of the proposal point out that there are no studies demonstrating the minimum effective size for a landscape linkage. We

FIGURE 7.1 The Klamath Corridors Proposal.

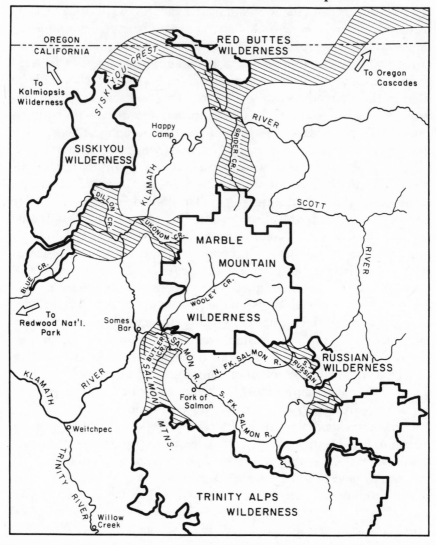

must remember, however, that such studies are probably not feasible because of the many factors that must be considered and because adequate controls are not available. To suggest that we should not act unless we have conclusive data is to divorce the goal of conservation from the discipline of biology. In this case it would mean foreclosing options that may prove critical to maintaining biodiversity in the Klamath National Forest. While monitoring should be done to evaluate how strategies perform, we should not hesitate to use what knowledge we have to design and implement management strategies.

A MODEL FOR THE FUTURE?

The Klamath Forest Alliance believes the Klamath Corridors Proposal applies the best available scientific thinking on how to maintain biodiversity. While there is no consensus on specific requirements, we can say with assurance that connectivity is superior to isolation, that continuity is preferable to fragmentation, and, with respect to corridors, that larger is better than smaller. The Klamath Forest Alliance, with support from other conservation organizations, has taken action in federal court to block logging in one of the Klamath Corridors while advocating study of the need for biological connectivity in the land management plan being prepared for the Klamath National Forest. We are also working to extend the proposal to encompass the entire Klamath Mountain Province. And, finally, we are exploring options for landscape linkages to reestablish native forest habitat connectivity between the Klamath Mountain Province and both the Oregon Cascades to the northeast and the Sierra Nevada to the southeast.

Within the native forests of the West Coast there remain promising options for ensuring maintenance of biodiversity through retention of natural habitat connectivity. In another decade these options will be gone. If we are to apply the strategies pioneered by Larry Harris and Reed Noss, the time to act is now. Biologists in the agencies and in academia, with the support of an informed public, should develop landscape linkage proposals and advocate their adoption. We think the Klamath Corridors Proposal could well serve as a model for such efforts.

EDITOR'S NOTE: On September 13, 1990, a federal appeals court ruled in favor of the Marble Mountain Audubon Society, which had challenged USDA Forest Service timber sales in the Grider Creek drainage of the Klamath National Forest. In the landmark decision, the court ruled that federal agencies must consider an area's importance as a "biological corridor" linking wilderness zones before permitting logging. This ruling represents the first federal court recognition of the significance of biological corridors.

REFERENCES

Atzet, T., and D. L. Wheeler, 1984. Preliminary plant associations of the Siskiyou Mountain Province. Portland: USDA Forest Service, Pacific Northwest Region.

California Department of Fish and Game. 1987. Forest mammal survey and inventory. Sacramento: Calif. Dept. of Fish & Game.

National Council on Gene Resources. 1982. Anadromous salmonid genetic resources: an assessment plan for California. Berkeley: National Council on Gene Resources.

Newmark, W. D. 1987. A land-bridge island perspective on mammalian extinctions in western North American parks. *Nature* 385:430–432.

Perry, D. A., M. P. Amaranthus, J. G. Borchers, S. L. Borchers, and R. E. Brainerd. 1989. Bootstrapping in ecosystems. *Bioscience* 39:230–237.

Sawyer, J. O., D. A. Thornburgh, and W. F. Bowman. 1970. Extension of the range of *Abies lasiocarpa* into California. *Madrono* 20(8):413–415.

Thomas, J. W., et al. 1990. A conservation strategy for the spotted owl. Portland: USDA Forest Service et al.

8 Faunal Movement Corridors in Florida

Larry D. Harris and Kevin Atkins

OVER 125 YEARS AGO the eminent European ecologist de Candolle made a remarkable observation: "The breakup of a large landmass into smaller units would necessarily lead to the extinction or local extermination of one or more species and the differential preservation of others" (quoted in Browne 1983). Curtis (1956) was perhaps the first to illustrate habitat fragmentation with a southern Wisconsin situation analysis nearly thirty-five years ago, and others have since described the phenomenon in greater detail (Burgess and Sharpe 1981; Harris 1984; Saunders et al. 1987). Increasing numbers of scientists are reaching the conclusion that "habitat fragmentation is the most serious threat to biological diversity and is the primary cause of the present extinction crisis" (Wilcox and Murphy 1985:884).

There are two obvious approaches to reducing the effects of habitat fragmentation: Increase the size of nearby conservation areas until they conjoin, or conserve or restore habitat corridors between the areas. Nearly thirty years ago, Preston (1962:427) concluded that:

> If the major part of the State, for instance, is given over to a complete disclimax, whether urbanization or mining or agriculture, the preserved area becomes an "isolate" or an approximation thereto, and the number of species that can be accommodated must apparently fall to some much lower level. . . . The only remedy is to prevent the area from becoming an isolate by keeping open a continuous corridor with other preserved areas. Even then there will be a nationwide impoverishment due to mere reduction of area as natural conditions are replaced by unnatural ones.

But the relationship of ecological principle to on-the-ground management is not always obvious, and it is not uncommon for decision
makers to dismiss such statements as empty theory. We believe, however, that the loss of biological diversity in rapid-growth states is now
so critical that any further dismissal of the need to implement a system
of wildlife movement corridors constitutes a form of social and administrative neglect.

This chapter reviews the historical applications of animal movement
corridors in the conservation of wildlife resources and suggests that
the principle of these corridors must now be codified and adopted as a
central approach to the preservation of threatened biological diversity.
In rapidly developing areas such as Florida, each passing year will
make it increasingly expensive—and increasingly ineffective—to apply the movement corridor strategy on a case-by-case basis. We suggest that the strategy be adopted as an essential complement to other
approaches that have traditionally dominated the wildlife conservation
movement.

CONSEQUENCES OF HABITAT FRAGMENTATION

Conservation biology research has progressed far, and the demonstrated effects of habitat fragmentation can now be assigned to several
classes. Although these effects have been observed widely throughout
the world, we describe the situation in Florida as an example of the
phenomenon in general.

First, the loss of habitat-interior species, those that occur only in the
interior of relatively large tracts, is an immediate diagnostic of the
habitat fragmentation issue (see Harris 1989; Harris and Silva-Lopez
1990). The sharp-shinned hawk (*Accipiter striatus*), Cooper's hawk (*A.
cooperii*), Swainson's warbler (*Limnothlypis swainsonii*), and the red-
cockaded woodpecker (*Picoides borealis*) are examples of species that
require relatively large tracts of specific habitat types in order to
reproduce successfully and maintain a viable population. Although
this type of effect is most commonly reported for bird species, more
detailed research will surely confirm that it is valid for other taxa
as well.

Second, when the remaining habitat fragments occur in a human-dominated landscape matrix, wide-ranging species are more likely to perish as they attempt to move between remaining patches. Wolves (*Canis* spp.), black bears (*Ursus americanus*), and panthers (*Felis concolor*) are examples of species that readily adapt to moderately managed landscapes when they can move between protected areas without molestation or death. But in heavily settled areas, such as Florida, these species are now either extirpated or critically threatened by habitat fragmentation at the regional and statewide scale. Indeed, boat and automobile traffic has now emerged as the number one habitat-fragmenting force and the primary source of human-related mortality for all of Florida's large threatened and endangered species (Harris and Frederick 1990). How, then, can the traditional approach of designating a series of small, isolated, protected areas conserve America's largest and most wide-ranging inshore/inland mammal, the West Indian manatee (*Trichechus manatus*), when the predominant threat to its survival is boat collisions during the course of migrating to reserves or foraging outside them?

Third, consequences of random variation in demographic and genetic variables are quick to threaten those species that are sufficiently sedentary to remain within protected but isolated fragments of habitat. For example, the demographic and genetic consequences of inbreeding are now known to pose serious threats to preservation of the endangered Florida panther (*F. c. coryi*). (See USFWS 1987; Seal and Lacy 1989.)

Fourth, increasing dominance of the wildlife community by alien and "weedy" species gradually transforms regionally distinctive faunas into a globally homogeneous array of species that are tolerant of humans and their domestic animals. Thus, despite the high number of extinct, endangered, and highly threatened native species, the overall increase in biological diversity (taken literally) is rapidly increasing in species by any measure. While only a few species of plants have been lost to extinction, for example, the native flora is complemented by literally hundreds of alien species. About one-third of peninsular Florida's vegetation now consists of alien species (Dr. R. Wunderlin, pers. com.). Similarly, whereas fewer than a dozen species of native vertebrates have been lost to extinction during the last 500 years, we have gained nearly 100 species of free-ranging exotic vertebrates (including freshwater fish).

Fifth, greatly altered ecological processes that govern existence and evolution in native ecosystems are the obvious end result of all these changes in animal community characteristics. In the absence of native large predators, hunters, and trappers, and with fragmentation of formerly expansive tracts of closed-canopy habitat, the proliferation and depredations of even a single species, such as the raccoon (*Procyon lotor*), now can threaten game birds, gopher tortoises (*Gopherus polyphemus*), marine turtles, and more than 100 other native species of vertebrates that must nest on or very near the surface of the ground. In addition, raccoons carry rabies, a disease that threatens not only humans but also Florida panthers and other endangered species.

GRAVITY OF THE ISSUE

Reviewing a single group—Florida's large mammals—dramatizes the gravity and urgency of the faunal conservation challenge in Florida. Until 200 years ago, Florida was inhabited by eleven species of native mammal larger than 5 kilograms in size: bison (*Bison bison*), manatee, black bear, monk seal (*Monachus monachus*), white-tailed deer (*Odocoileus virginianus*), key deer (*O. v. clavium*), Florida panther, red wolf (*Canis rufus*), bobcat (*Lynx rufus*), otter (*Lutra canadensis*), and raccoon. Of these, the monk seal, bison, and red wolf are globally or locally extinct; the manatee, Florida panther, and key deer are federally listed as endangered; and the black bear, bobcat, and otter are either listed as threatened by the state of Florida or listed by the Convention on International Trade in Endangered Species (CITES). Thus only two species of native large mammal in Florida, the white-tailed deer and the raccoon, are doing well.

But the native species have been substituted for by numerous species of smaller, more adaptable, and generally more omnivorous aliens and exotics. Cattle graze in place of bison; coyote (*Canis latrans*), red fox (*Vulpes fulva*), and feral domestic dogs (*Canis familiaris*) have replaced red wolves as predators; feral hogs (*Sus scrofa*) are free-ranging throughout the state and rhesus monkeys (*Macaca mulatta*) are locally abundant.

Agencies or decision makers that emphasize wildlife diversity and abundance—without putting strong emphasis on preserving native faunal assemblages—contribute to regional and global homogenization of biological diversity just as surely as do those who encourage the introduction of exotic species (ornamental plant distributors, for instance, and biological control specialists). Until we succeed in establishing that conserving animals in their native habitat is more important than simply maintaining species per se, the public will continue to be content with "unnatural" tourist attractions such as the rhesus monkey population at Silver Springs in Florida's Ocala National Forest (Wolfe and Peters 1987), rather than demanding that they be conserved in their native country of India.

But most troubling of all, Americans have been schooled to believe that increasing the abundance of any species—be it raccoon, robin (*Turdus migratorius*), or ring-necked pheasant (*Phasianus colchicus*)—is not only acceptable wildlife conservation but a resource management ideal worthy of praise. If we are to be true to the cause of conserving biological diversity, we must reduce our emphasis on wildlife per se, which includes all free-ranging species, and increase our focus on the native faunal assemblages that distinguish one region from another.

FAUNAL MOVEMENT CORRIDORS: BLESSING OR BANE?

The importance of movement corridors will not be apparent to those who neither sense the gravity of the biodiversity crisis nor identify with the goal of conserving native fauna and flora as unique assemblages of species. A key impediment to progress is that those who wish to defend the status quo and those who fail to appreciate the inadequacy of present conservation paradigms see no need for new approaches. Many planners, developers, and legislators fail to understand that plants and animals need to move—and that simply designating isolated protected areas that are surrounded by human occupation cannot meet this need. Inadvertently, critics have created the impression that movement corridors are unnecessary, ineffective, akin to pushing panthers through culverts. With a cavalier disregard for the reality of the connectivity of native landscapes prior to human settlement, critics of movement corridors claim that maintaining or reestablishing natural linkages could be ecologically damaging because they could facilitate the spread of predators, parasites, and diseases. Thus decision makers have been presented a simple black-and-white dichotomy: Corridors are either the solution to all the problems cited here or the bane of future conservation efforts. Both extremes are far from the truth.

Wildlife movement corridors will not solve all of these problems any more than would the establishment of new isolated parks, the passage of new laws, the listing of new endangered species, or a doubling of law enforcement staff. But given that Florida already has a system of protected areas, given that the raccoon is already listed as a game species and the panther is listed as endangered, and given that wildlife law enforcement is active and effective, our next major step must be implementation of an integrated statewide system of landscape linkages and corridors that interconnect the largest of the protected areas and allow for movement between other critically isolated populations.

The crucial questions are not whether corridors are all good or all bad, not whether they are the sole answer to the present biodiversity crisis, or even whether there are costs and liabilities. Rather, the key question is whether an integrated system of protected and managed natural habitats will be less discriminatory against our rare and endangered native species and also less supportive of alien, exotic, and pest species. Will such an interconnected system of habitats be superior for more natural assemblages of native species and more natural levels of ecological processes, such as competition, predation, and parasitism, than a disjunct system of isolated preserves? The consensus of leading conservationists is a resounding yes.

THE LONG HISTORY OF CORRIDOR USE

Critics of faunal corridors have suggested that they are simply an untested application of unproved island biogeography theory. The wildlife management record attests otherwise, however, for corridors have been widely used for over fifty years. The most ubiquitous use of wildlife movement corridors occurred in the early decades of this century with regard to the management of game species in rapidly developing agricultural landscapes of the Midwest (Edminster 1938; Davison 1939, 1941; Petrides 1942; Dambach 1945, 1948). This was especially important for the game birds (Sumner 1936; Harmon and Damon 1947; Gehrken 1975) and the squirrels (*Sciurus* spp.: Allen 1943; Baumgartner 1943; McElfresh et al. 1980; Flyger and Gates 1982; Dickson and Huntley 1987). When used in this context, "travel corridors or lanes refer to wildlife cover, usually linear, that offers a safe roadway or route from one habitat to another" (Johnson and Beck 1988:314).

When used in the forestry context, corridors have been defined as "a strip of timber left at the time of harvest. Its main purpose is ease of animal movement across areas that are at first too open (new clearcutting) or later too dense (such as sapling stands)" (Landers 1985:4). Fishery managers had developed techniques to maintain movement opportunities for migratory fish around major dams and hydroelectric generators, and big-game biologists were quick to develop similar techniques for ungulate movement in the presence of high-speed interstate highways and related transport systems (Klein 1971, 1980; Reed 1981; Ward 1982; Pedevillano and Wright 1987; McDonald 1988).

Although they relied less upon technological solutions and more upon simple retention of natural movement corridors, large-mammal biologists demanded implementation of corridor systems throughout the world. Virunga National Park in Zaire, for example, perhaps the first national park in Africa, is complemented by movement corridors, as are several other East African parks. The same techniques were recommended for tigers (*Panthera tigris*: Seidensticker and Hai 1983; Seidensticker 1989), Asian elephants (*Elephas maximus*: Rudran et al. 1980), giant pandas (*Ailuropoda melanoleuca*: Schaller et al. 1985), black bears (Pelton 1986), grizzly bears (*Ursus horribilus*: Picton 1988), Florida panthers (Maehr 1990), and for recovery of the several endangered cat species of the southwestern United States (Harwell and Siminski 1988).

Urbanization has forced the need to design environmental corridors into human-dominated environments, and urban and regional planners have long seen the necessity of habitat corridors to connect otherwise isolated parks. (See Lewis 1964; Gill and Barnett 1973; Davis and Glick 1978; Douglas 1983; Poynton and Roberts 1985.) Because roads have come to represent such major barriers to movement in developed-world landscapes, the sale of commercially produced underpasses and related technologies is now widespread (see especially Langton 1989).

HOW MUCH CAN BE DONE?

It is essential to bear in mind the issue of scale when deliberating on the necessary dimensions of corridors and landscape linkages. If they are to mimic natural landscape features, the width of corridors should be gauged in units of the landscape, not simple, human-imposed numbers. And while the value of a natural vegetation strip along a creek may be extremely great for small creatures, it may be totally useless for other species, such as large vertebrates, or for the movement of entire faunas and floras. Perhaps the following guidelines will be useful to planners and decision makers:

- When the movement of individual animals is being considered, when much is known about their behavior, and when the corridor is expected to function in terms of weeks or months, the appropriate corridor width might be measured in tens of meters (that is, 30–300 feet) for individual Florida scrub jays or red-cockaded woodpeckers.
- When the movement of an entire species is being considered, when much is known about its biology, and when the corridor is expected to function in terms of years, the appropriate corridor width might be measured in hundreds of meters (that is, 300–3,000 feet) for iterative population interchange.
- When the movement of entire assemblages of species is being considered, when little is known of their biology, and when the faunal dispersal corridor is expected to function over decades or centuries, the appropriate width should be measured in kilometers (greater than 3,000 feet for northward dispersal because of climate change).

THE TREND TOWARD CORRIDOR SYSTEMS

The combined efforts of government agencies and nongovernmental conservation organizations have kept Florida at the forefront of conservation programming for nearly 100 years. While protecting the environment has now become politically important throughout the country, this is especially true in Florida. As illustrated in Table 8.1, a sequential development of science, laws, and conservation activities has set the stage for serious wildlife corridor programming. This progression can be described in phases.

Phase I: Early 1900s. At the turn of the century, natural history was the underlying science, the regulation of human behavior directed specifically at wildlife was the predominant approach to conservation, and the Lacey Act and Migratory Bird Treaty Act were principal legal mechanisms. Since large expanses of unsettled lands were still available for designation as conservation areas, Pelican Island, the country's first national wildlife refuge, and Ocala National Forest, the first national forest in the eastern United States, were created by presidential designation.

Phase II: Mid-1900s. By mid-century, ecology was a recognized science, legislation such as the Pittman-Robertson (P-R) Act was aimed at wildlife habitat and not just the species themselves, and the powerful state and federal regulations were complemented by habitat acquisition and management. Money from the P-R Act allowed individual states to acquire conservation estates. Florida purchased its first wildlife management areas by mid-century (Corbett and Webb WMAs, for example). The 1960s ushered in a suite of federal legislation aimed at regulating human behavior toward the entire environment, not just toward wildlife per se. The Multiple Use Sustained Yield Act (MUSY), the National Environmental Protection Act (NEPA), and several versions of the Endangered Species Act (ESA) all were critical to conservation of biodiversity in Florida. At the ground level, blockage of several development schemes expressed the changing attitudes and values of Florida's citizens.

Phase III: 1970s. The seventies were characterized by yet newer approaches and levels of conservation activity. Because pesticides were endangering many species, detergents were polluting the water, and automobile fumes were fouling the air, unguided human population growth simply had to be better managed. State land acquisition emerged as the necessary complement to federal and state regulation. Systems ecology had become an important subdiscipline and a flurry of systems analyses were directed at Florida's fast-disappearing natural ecosystems. Four 1972 laws provided legal instruments for the new conservation movement. The Florida Water Resources Act created five Water Management Districts (WMDs) that had the effect of creating a new tier of regional regulatory and land acquisition authority; ultimately, these WMDs were given the all-important power of taxation.

While the Environmental Land and Water Management Act and the Comprehensive Planning Act mandated state-level planning and critical evaluation of developments of regional impact (DRIs), the Land Conservation Act authorized purchase of environmentally endangered lands (EELs).

Phase IV: 1980s. The eighties brought further revamping of policies and approaches. Landsat photography and Geographic Information Systems (GIS) illustrated the gravity of the habitat fragmentation issue and served as key analytic technologies. Habitat mitigation, reclamation, and restoration became critical approaches to conservation programming. There was a widespread realization that inadequately controlled growth and development had proceeded too far. Restoration of the Everglades, dechannelizing the Kissimmee River, removal of Rodman Lake and restoration of the Oklawaha River—all are examples of efforts to not only halt but reverse the destructive trends. Expanded budgeting programs aimed at wetlands protection and habitat acquisition allowed the integration of science, technology, and advanced conservation philosophies on the ground.

Phase V: 1990s. Despite the many legislative and programming advances described here, it remained obvious that piecemeal conservation programming would not suffice in Florida in the twenty-first century. Although public sentiment strongly favored protection and restoration of wildlife corridors and greenways—and strongly opposed further urban sprawl and habitat fragmentation—outdated conservation paradigms that hinged on the purchase of small, isolated parks for regional solutions still persisted in 1990. But a national and international consensus had emerged among conservation biologists that small, isolated parks were simply not able to protect Florida's natural systems or their critical ecological functions. Pressing questions—such as how local governments could achieve compliance with the planning laws and how the endangerment of wide-ranging species such as manatees and Florida panthers could be reversed by isolated preserves—became increasingly critical in the minds of Florida citizens. While landscape ecology and restoration ecology provided scientific underpinning, the concepts of mitigation banking, land swapping, and debt swapping evolved as new conservation technologies. A land use planning alternative written into the Florida statutes provided incentive for the private sector to focus on Florida Quality Developments (FQDs) and aid in the conservation, restoration, and connection of natural upland and wetland habitat.

PRESERVATION 2000

In 1989, Florida's Governor Martinez appointed Nathaniel P. Reed to head the Commission on the Future of Florida's Environment. In April 1990 the commission released its report finding that natural systems were critical to Florida's $183 billion economy—but noting that funding of only 2.5 percent of the state's annual budget for environmental programming was hardly a fair share. The commission concluded that "land acquisition should become a cornerstone of local, regional, and state land use planning . . . , and preservation of large tracts of ecologically important lands, greenways, wetlands, and water recharge areas" was necessary. The commission recommended that "the 1990 Florida legislature enact law to significantly enhance the funding of the state's environmental land acquisition and restoration programs by authorizing a bonding program to raise $300 million per year for 10 years." This $3 billion land acquisition and management program, now referred to as Preservation 2000, is surely one of the largest single-focus conservation projects in the world.

Because the Preservation 2000 program is funded primarily by the sale of bonds, it is now clear that the future citizens of Florida are the ones who will pay for Florida's future. Incurring this level of debt at a time when America's debt load is a major political issue causes anxiety among some; it may help allay their fears to know that the decisions of today will, in fact, contribute significantly to Florida's biological diversity tomorrow. It is in this vein that we wish to clarify what is meant by faunal movement corridors and the role they might play in the Preservation 2000 program.

We define a faunal dispersal corridor as a naturally occurring or restored linear landscape feature that connects two or more larger tracts of essentially similar habitat and functions as either a movement route for individuals or an avenue for the spread of genes or other natural ecological processes across the landscape. It is critical, however, to bear certain caveats in mind. First, conserved or created vegetation strips that consist of alien species or communities do not

TABLE 8.1 Florida's Progression Toward Biodiversity Conservation

Conservation Concept	*Underlying Science*
Early 1990s	
• Federal laws	• Natural history
• Federal reserve designation	
• State regulation	
Mid-1900s	
• Federal regulation and acquisition	• Ecology
• State regulation and acquisition	• Management sciences
1970s	
• Federal regulation and acquisition	• Systems ecology
• Accelerated state acquisition	• Water resource management
• Regional impact assessment	• Land use planning
• Land reclamation	
1980s	
• Federal and state regulation and acquisition	• Systems ecology
• Regional regulation and acquisition	• River and wetlands systems
• Comprehensive growth management planning	• Landsat habitat mapping
• Mitigation and restoration	• Restoration ecology
1990s	
• Federal, state, and regional regulation and acquisition	• Restoration ecology
• Local regulation and acquisition	• Geographic information systems
• Growth management compliance	• Landscape ecology
• Mitigation banking	
• Land swaps/debt swaps	
• Incentives	
• Landscape linkages	

Laws or Instruments	Trend in Acquisition and Implementation
• Executive proclamation • Federal legislation	• Pelican Island NWR • Ocala NF
• Accelerated federal acquisition programs • Pittman-Robertson Act • Multiple Use and Sustained Yield Act • National Environmental Policy Act	• Everglades NP • St. Marks NWR • Corbett WMA
• Endangered Species Act • Federal Clean Water Act • Florida Environmental Land and Water Management Act • Environmental impact statements • Development of regional impact process • Environmentally Endangered Lands Acquisition Program	• Many state and national preserves (Big Cypress NP) • "Biodiversity elements" protection (Florida scrub) • Everglades conservation areas consolidation • Mandatory wetlands reclamation for phosphate mining
• Conservation and recreation lands acquisition • Henderson Wetland Protection Act • Florida Growth Management Act • Water quality legislation • Save Our Rivers/Save Our Coasts acquisitions	• Kissimmee–Okeechobee–Everglades Regional System restoration • Okefenokee NWR–Osceola NF linkage • Cross-Florida Barge Canal deauthorization
• Comprehensive plans and performance standards • Florida Quality Development Program • Upland habitat protection rules • Recovery plans for threatened and endangered species • Florida Preservation 2000 Act	• Kissimmee River Corridor restoration • Cross-Florida "Greenway Park" • Florida panther corridors and underpass • Big Bend Gulf Coastal Corridor

necessarily qualify as acceptable faunal dispersal corridors. Road rights-of-way across natural wetlands, or canals and levees across the Everglades, or powerlines running through a state preserve create important linear corridors, for example, but they also cause an alien habitat to bisect native community types. Although these corridors will function as wildlife dispersal routes, the dispersing species are likely to be alien or weedy species and the net effect may not be desirable. By way of contrast, fencerows or highway rights-of-way consisting of retained or restored native vegetation that interconnect larger tracts of native habitat do qualify as faunal corridors.

Second, even though faunal dispersal corridors generally consist of linear strips, not all linear habitats constitute faunal corridors. This is because many linear habitats neither come from nor lead to larger tracts of similar habitat. They may simply be long, narrow, habitat fragments—precisely what the corridor principle is meant to alleviate. Long but narrow strips of habitat may serve as ecological traps for species that use them just as surely as short and narrow ones can. It is becoming increasingly clear that small or linear parks and protected areas are prone to dominance by edge species and do not truly preserve the native fauna and flora. Most regions do not need more small fragments of predominantly edge habitat in the name of faunal conservation, be they linear, circular, or any other shape.

Third, greenline parks and other linear habitats such as restored rights-of-way may appear to serve as faunal dispersal corridors. But if the greenline parks are too narrow or too alien, they may represent more of a liability than a functional benefit. For example, protection of a narrow riparian wood along a channelized urban creek may offer numerous benefits for humans and wildlife alike. But unless it meets specific standards, it should not be advocated as a faunal dispersal corridor. Powerlines or cleared fencerows that bisect natural vegetation communities should not be sanctioned or supported in the name of faunal corridors.

Like most regions, Florida needs many more protected areas and greenspaces, some of which must be small. But the purchase of more small, isolated areas cannot be assigned the same priority as the implementation of a corridor system that builds toward an integrated system of habitats for the maintenance of native ecological processes as well as native biodiversity. If the maintenance of natural ecological processes—pollination, propagule and gene dispersal, predation, competition, burning, and the like—are important to the conservation of native fauna and flora, then Preservation 2000 must focus on the restoration of an integrated, statewide habitat system that perpetuates these natural processes. Because of the predicted increases in temperature and sea level that will result from the greenhouse effect, the interconnected system must encompass large expanses if it is to facilitate the movement of entire faunas and floras across a human-dominated landscape.

REFERENCES

Allen, D. 1943. *Michigan fox squirrel management*. Game Division Publ. 100. Lansing: Michigan Department of Conservation.

Baumgartner, L. 1943. Fox squirrels in Ohio. *J. Wildl. Management* 7:193–202.

Browne, J. 1983. *The secular ark: studies in the history of biogeography*. New Haven: Yale University Press.

Burgess, R. L., and D. M. Sharpe (eds.). 1981. *Forest island dynamics in man-dominated landscapes*. New York: Springer-Verlag.

Curtis, J. 1956. The modification of mid-latitude grasslands and forests by man. Pp. 721–736 in *Man's role in changing the face of the earth* (W. Thomas Jr., (ed.). Chicago: University of Chicago Press.

Dambach, C. A. 1945. Some biological and economic aspects of field border management. Pp. 169–184 *Trans. N. Am. Wildl. Conf.* Field Border Management 10.

————. 1948. *A study of the ecology and economic value of crop field borders*. Biological Science Series no. 2, Graduate School Studies. Columbus: Ohio State University.

Davis, A., and T. Glick. 1978. Urban ecosystems and island biogeography. *Environ. Conserv.* 5(4):299–304.

Davison, V. E. 1939. *Protecting field borders*. Leaflet 188. Washington: U.S. Department of Agriculture.

————. 1941. Wildlife borders—an innovation in farm management. *J. Wildl. Management* 5:390–394.

Dickson, J., and J. Huntley. 1987. Riparian zones and wildlife in southern forests: the problem and squirrel relationships. In *Managing southern forests for wildlife and fish* (J. Dickson and O. Maughan eds.) USDA Forest Service Gen. Tech. Rept. SO-65.

Douglas, I. 1983. *The urban environment*. London: Edward Arnold.

Edminster, F. C. 1938. Woody vegetation for fence rows. *Soil Conserv.* 4(4):99–101.

Flyger, V., and J. Gates. 1982. Fox and gray squirrels. In *Wild mammals of North America* (J. Chapman and G. Feldhamel, eds.). Baltimore: Johns Hopkins University Press.

Gehrken, G. A. 1975. Travel corridor technique of wild turkey management. In *Proceedings of the National Wild Turkey Symposium* (L. K. Hall, ed.). Austin: Texas Chapter of the Wildlife Society.

Gill, D., and P. Barnett. 1973. *Nature in the urban landscape: a study of city ecosystems.* Baltimore: York.

Harmon, W., and D. Damon. 1947. Hedgerow management—bobwhite's ally. *Outdoor Nebraska* 23(1).

Harris, L. D. 1984. *The fragmented forest: island biogeography theory and the preservation of biotic diversity.* Chicago: University of Chicago Press.

————. 1989. The faunal significance of fragmentation of southeastern bottomland forests. In *Proceedings of the forested wetlands of the southern United States* (D. Hook and R. Lea, eds.). USDA Forest Service Gen. Tech. Rept. SE-50.

Harris, L. D., and W. Cropper. 1990. Between the devil and the deep blue sea: implications of climate change for wildlife in a southeastern growth state. In *Climate change and conservation* (R. Peters and T. Lovejoy, eds.). New Haven: Yale University Press.

Harris, L. D., and P. Frederick. 1990. The role of the Endangered Species Act in the conservation of biological diversity: an assessment. In *Integrated Environmental Management* (J. Cairns, Jr., and T. Crawford, eds.). Chelsea, Mich.: Lewis.

Harris, L. D., and J. Scheck. 1990. From implications to applications: the dispersal corridor principle applied to the conservation of biological diversity. In *Nature conservation: the role of corridors* (R. Hobbs and D. Saunders, eds.). Chipping Norton, NSW: Surrey Beatty & Sons.

Harris, L. D., and G. Silva-Lopez. 1990. Forest fragmentation and the conservation of biological diversity. In *Conservation biology: the theory and practice of nature conservation, preservation and management* (P. Fielder and S. Jain, eds.). New York: Chapman and Hall.

Harwell, G., and D. Siminski. 1988. Listed cats of Arizona and Texas recovery plan. U.S. Fish and Wildlife Service, Albuquerque.

Johnson, R. J., and M. M. Beck. 1988. Influences of shelterbelt on wildlife management and biology. *Agriculture, Ecosystems, and Environment* 22/23:301–335.

Klein, D. R. 1971. Reaction of reindeer to obstructions and disturbances. *Science* 173(3995):393–398.

————. 1980. Reactions of caribou and reindeer to obstructions—a reassessment. Pp. 519–527 in *Proceedings of the Reindeer/Caribou Symposium*, vol. 2 (E. Reimers, E. Gaare, and S. Skjenneberg, eds.). Roros, Norway.

Landers, J. 1985. *Integrating wildlife and timber management in southern pine forests.* Forest Management Guidelines no. 8. Bainbridge, Ga.: International Paper Company.

Langton, T. (ed). 1989. *Amphibians and roads.* Dedfordshire, England: ACO Polymer Products.

Lewis, P. H. Jr. 1964. Quality corridors for Wisconsin. *Landscape Architecture* 54(2):100–107.

McDonald, M. 1988. Glenn Highway Moose Monitoring Study progress report. Unpublished agency report submitted to Alaska Department of Transportation and Pub. Facilities by Alaska Department Fish and Game, Juneau.

McElfresh, R., J. Inglis, and B. Brown. 1980. Gray squirrel usage of hardwood ravines within pine plantation. Pp. 79–89 in *Proceedings of the Louisiana State University Forestry Symposium.*

Maehr, D. 1990. The Florida panther and private lands. *Conserv. Biol.* 4:1–4.

Pedevillano, C., and R. G. Wright. 1987. The influence of visitors on mountain goat activities in Glacier National Park, Montana. *Biol. Conserv.* 39:1–11.

Pelton, M. 1986. Habitat needs of black bears in the east. In *Wilderness and natural areas in the eastern United States: a management challenge* (D. Kulhavy and R. Conner, eds.). Nacogdoches, Texas: Center for Applied Studies, School of Forestry, Stephen F. Austin State University.

Petrides, G. A. 1942. Relation of hedgerows in winter to wildlife in central New York. *J. Wildl. Management* 6(4):261–280.

Picton, H. 1988. A possible link between Yellowstone and Glacier grizzly bear populations. *International Conference on Bear Restoration and Management* 6:7–10.

Poynton, J., and D. Roberts. 1985. Urban open space planning in South Africa: a geographical perspective. *S. Africa J. Sci.* 81:33–37.

Preston, F. W. 1962. The canonical distribution of commonness and rarity. *Ecology* 43:185–215 and 410–432.

Reed, D. 1981. Mule deer behavior at a highway underpass exit. *J. Wildl. Management* 45:542–543.

Rudran, R., M. Jansen, and J. Seidensticker. 1980. Wildlife. In *Environmental assessment: accelerated Mahaweli development program. Vol. II: Terrestrial environment.* Colombo: Ministry of Mahaweli Development, Democratic Socialist Republic of Sri Lanka.

Saunders, D., G. Arnold, A. Burbidge, and A. Hopkins. 1987. *Nature conservation: the role of remnants of native vegetation.* Chipping Norton, NSW: Surrey Beatty & Sons.

Schaller, G., H. Jinchu, P. Wenshi, and Z. Jing. 1985. *The giant pandas of Wolong.* Chicago: University of Chicago Press.

Seal, U. S., and R. C. Lacy. 1989. Florida panther viability analysis and species survival plan. Captive Breeding Specialist Group, Species Survival Commission, International Union for the Conservation of Nature, in conjunction with U.S. Fish and Wildlife Service, Atlanta.

Seidensticker, J. 1989. Large carnivores and the consequences of habitat insularization: ecology and conservation of tigers in Indonesia and Bangladesh. Pp. 1–41 in *Cats of the world: biology, conservation and management* (S. Miller and D. Everett, eds.). Washington: National Wildlife Federation.

Seidensticker, J., and M. A. Hai. 1983. *The Sundarbans wildlife management plan: conservation in the Bangladesh coastal zone.* Gland, Switzerland: World Wildlife Fund.

Sumner, E. 1936. *A life history of the California quail, with recommendations for conservation and management.* Sacramento: California State Printing Office.

U.S. Fish and Wildlife Service. 1987. Florida panther (*Felis concolor coryi*) recovery plan. Prepared by the Florida Panther Interagency Committee for the U.S. Fish and Wildlife Service, Atlanta.

Ward, A. L. 1982. Mule deer behavior in relation to fencing and underpasses on Interstate 89 in Wyoming. *Transportation Research Record* 859:8–13.

Wilcox, B. A., and D. D. Murphy. 1985. Conservation strategy: the effects of fragmentation on extinction. *Am. Naturalist* 125:879–887.

Wolfe, L., and E. Peters. 1987. History of the freeranging rhesus monkeys (*Macaca mulatta*) of Silver Springs. *Florida Scientific* 50:234–244.

Discussion

The following discussion reflects the audience's concern with the continuing loss of biodiversity and the practical applications of corridor models.

TONY POVILITIS (HUMANE SOCIETY OF THE UNITED STATES): I would like to address my question to Dr. Harris. Given the fact that the forest patches and ecosystem fragmentations must result in the saturation of the species, do you expect that as time goes on there will be a relaxation resulting in further loss of species diversity and overall biodiversity? And if, by some stroke of luck, there were a rule that there will be no further loss of habitat in Florida, what results do you see in your crystal ball?

LARRY HARRIS: I think the wildlife profession needs to take a very serious look at its definition of habitat. I don't think we have a clue what habitat is and what it isn't. I know for a fact that a lot of developers are convincing legislatures that they are creating more habitat than they are destroying. I also know that people in Florida are bamboozled, and that they're not getting enough leadership from the professionals. We need to take the Clean Water Act and apply its precepts to maintaining balanced biological communities and balanced indigenous populations.

TONY POVILITIS: If there were a complete moratorium on, say, landscape development, do you believe that biodiversity legislation would be satisfactory from the point of view of ecologists?

LARRY HARRIS: No, not at all. I think we're doing a miserable job of managing our public lands and especially in states like Florida. I believe we just have to be much more aggressive at managing the lands we have. And I'm not antigrowth, either. If California can have 20 million people, why can't Florida?

TONY POVILITIS: Don't we send a double message to society as conservationists? Sending the message that we can have our cake and eat it too?

LARRY HARRIS: I think we *can*. I think a very rich state, like Florida, can afford it.

FRED SAMSON (FOREST SERVICE IN JUNEAU, ALASKA): First, I'd like to thank Dr. Soulé for his comments on the negative aspects of corridors; a number of us have published on those aspects for years. I'd also like to comment on your corridors model. As a person involved in getting these concepts off the drawing board and onto the ground, and trying to explain them to land managers, I find your model extremely complex. Are there any practical applications of your model in the field?

MICHAEL SOULÉ: I think where we are now in understanding corridors is where we were in 1962 in understanding island biogeography. Then there wasn't any theory. Not until the mid-sixties did we start to develop and test the theory, which led to an awareness in the academic and management communities of the importance of fragmentation, islands, and the concept of relaxation and loss of species diversity. I think we're still in the pre-1962 stage of understanding corridors. They've become a cause célèbre. Obviously fragmentation is the major issue, and corridors are one way of dealing with it. But just to retrofit the landscape with corridors, willy-nilly, before beginning to think about what we're doing is going to lead to some failures—some very costly public relations failures—and to the exacerbation of many problems. So I'm not saying we should have a moratorium on corridors. I think maybe 5 or 10 percent of funds being spent on corridors should go to research—behavioral research, ecological research, and some modeling research—so that, in parallel with the development of corridors, we develop a scientific basis for what we are doing.

UNIDENTIFIED SPEAKER: Just to follow up, we need to become comfortable with Dr. Noss's emphasis on endemics and connectivity as sort of the general practical application of the concept. It seems to me it's a very appropriate recommendation for land managers right now.

MICHAEL SOULÉ: Yes, it is. It's a matter of scale, too. It's good to have corridors linking patches—small patches that are close together and where you know, for example, that you're not introducing organisms from the eastern United States into the western United States. That kind of scale issue needs to be understood. So, on a local and somewhat regional scale, corridors are absolutely necessary.

JIM PISSOT (NATIONAL AUDUBON SOCIETY): I have a question for Dr. Harris. Most of us are familiar with your work in both Florida and the Pacific Northwest. The slides you've shown this afternoon attempt to link Florida's national forests, state forests, wildlife refuges, and so on. There seems to be an implicit assumption that if, indeed, you can link these large bodies of public land that are less subject to development pressures, then we will have a big rate of success. If you're making recommendations in South Florida for grants to prepare corridors between large landmasses, are you also making recommendations for management?

LARRY HARRIS: Yes, indeed. Take the spotted owl issue, for example. It's true that birds fly and it's true that they may be able to fly over clearcuts. But we do not have any evidence that I know of that spotted owls either could or would use corridors. I think it's a very short-sighted policy that designs a fragmented old-growth system for owls but doesn't consider a thousand other species. And it just seems intuitive to me, as Reed Noss pointed out, that if connectivity is the natural baseline from which we're working, and if fragmentation is the artificial, then I'm not lobbying for artificial corridors. I am lobbying, though, for retention and restoration of natural connectivity—on a small scale, as well.

DAN BAKER (NATIONAL PARK SERVICE): I am curious about the relationship between corridors that are designed with certain minimum proportions and the number of corridors that are connecting different patches. Also, are there any studies or models that address habitat network complexity? That is, the numbers of maximum corridors between the number of maximum patches to achieve certain kinds of movement potential. This, I think, may be one order in magnitude above the complexity you're dealing with, but I would like to know if there are any experiential bases for that, as well as theoretical ones.

MICHAEL SOULÉ: The short answer and probably the only answer I can give to that is no. We are still at the early stages of trying to figure out what are the important variables in a two-dimensional system, where one corridor links two habitat patches and one species tries to move along the corridor. When we talk about many species in a three-dimensional landscape, the complexity of the problem, as you can imagine, increases more than exponentially. It's the difference between the two-body problem and the three-body problem in gravitational physics; the first is tractable, the second is not.

UNIDENTIFIED SPEAKER: It seems to me this discussion today suggests "newness." But it sounds remarkably similar to me like the challenges of the 1930s to 1950s. I'm curious to know if the panel sees today's problems—whether procedural or theoretical—as any different than the problems of, say, thirty or forty years ago?

FELICE PACE: I think the significant difference between today and yesterday is that there are far fewer options. People say there are still lots of options in the West. But when I look at the ground level, where I live, I don't see very many options, and I see even fewer for the rest of the country.

III

Reintegrating Humans and Nature

Introduction

ALLEN COOPERRIDER

PREVIOUS PARTS of this book have dealt with the mechanics of conserving biodiversity and counteracting habitat fragmentation. In Part III, we turn to the changes in human attitudes and behavior necessary to effect these changes.

Miller et al. (1989) point out that the immediate causes of loss of biological diversity are clear—biological resources are degraded, and species and habitats are fragmented or lost. The root causes of the loss are less obvious, but sooner or later they must be addressed. Furthermore, these root causes, which derive from human demands for commodities are inseparable from the other major problems of our day such as poverty, social injustice, and war. Even the current savings and loan debacle, for example, is exacerbating the loss of biodiversity (Bunch 1990). Therefore, the problem of addressing biodiversity cannot be addressed in isolation if long-term solutions are to be found.

Conservation of biodiversity will be neither simple nor painless. And even when solutions have been found, lack of consensus or willpower may prevent their implementation. Conservation biologists may be able to describe practical procedures to conserve a species or ecosystem, but their recommendations are not sufficient. An interagency group of scientists recently developed a well-designed and scientifically sound strategy for the recovery of the spotted owl (Thomas 1990); acceptance and implementation of their recommendations have not been the inevitable result (Anon. 1990).

Loss of biodiversity is thus a crisis of character and culture—not just of conservation—as Wendell Berry has pointed out in other contexts (Berry 1977). Preservation of biological diversity will require not only a wholesale change in the way we treat our wildlands and wildlife, but

141

also in the way we live and do business. And these changes will only come about if we change the way we think about ourselves, about nature, and about the relationship between the two.

Integration of human activities with conservation of biodiversity is arguably one of the greatest challenges facing us today. Yet anthropologists and historians tell us that humans or humanlike animals have lived on the earth for thousands of years with minimal impact on biodiversity until recently. This suggests that our ability to coexist with fellow animals and plants—to be *integrated*—has gone seriously awry. Examination of this dis-integration of humans from their environment may provide insight into what is needed for re-integration.

A BRIEF HISTORY OF DISINTEGRATION

Humans have been on earth for over 100,000 years. Most early hunting and food-gathering societies probably did not kill more game or harvest more food than could be replenished by annual increase under normal ecological conditions (Narr 1956). Even the early primitive agricultural cultures, although they caused some local destruction, probably did it at a scale in space and time that allowed recovery.

Changes in Culture. With the beginning of civilization, however, the opportunities to cause significant declines in biodiversity increased dramatically. The rise of civilization was accomplished and sustained by the development of powerful and well-organized states with a drive to territorial expansion, by commerce in bulk and to distant places, by monetary economy, and by the growth of cities. The industrial revolution—sustained by the exponential expansion in the use of fossil fuel—significantly increased humans' ability to drastically alter the environment to the detriment of native biodiversity.

These events have been magnified and intensified in the last fifty years. Since the end of World War II, we have experienced a doubling in the world's population and in the use of fossil fuels—with concomitant growth in cities and suburban areas, accelerated clearing of forests, increased appropriation of water for industrial and agricultural use, and intensified use of farmland and grazing lands. Furthermore, the development of the chemical industry has created a vast array of new threats to humans and our ecosystems in the form of such modern "miracles" as pesticides, preservatives, and plastics.

Changes in Character. These changes in culture were accompanied by parallel changes in people's view of the world—that is, changes in human character. Life in cities separated people from direct association with the production of their food as well as from wildlife and wilderness. People became spiritually as well as physically removed from the land. As populations and institutions grew and human migrations increased, many people became increasingly detached from their communities. Even though people lived in settlements, they began to lose a sense of bonding with the place where they lived and the people of that place, a trend that continues to this day. Accompanying this separation was a loss of historical perspective. With limited contacts between generations occupying the same places, people's sense of history became limited. Dasmann (1985) has pointed out that people who feel continuity with the past and future and an attachment to place are much more likely to be good stewards of the land.

Religious thought that encourages a worldview of nature dominated by humans has played a significant role in supporting this attitude of separation (Watts 1958; White 1967). And the discovery and exploitation of fossil fuels (and later nuclear energy) has reinforced a feeling that humankind is all-powerful. Furthermore, monetary economies and capitalism have allowed humans to make decisions with devastating consequences to other people and other landscapes without personally observing or feeling the effects. The coining of money, for example, allowed a capitalist in Europe to invest in slave-operated enterprises in distant lands without the daily reminder of the suffering of people and land caused by such operations.

Moreover, the influence of modern science has played a major role in alienating humans from nature. Beginning with Descartes, science has emphasized reductionism—single factors are studied in isolation by attempting to keep all other variables constant. This has been a useful approach, but carried to extremes it has reinforced a mode of thinking that the world can be understood as a set of isolated parts rather than as an integrated whole. Reductionist thought not only reinforces the concept of humans separate from nature, but it encourages a human perspective that all of life can be compartmentalized.

THE FRAGMENTED GENERATION

This fragmentation of human thought and human activity is central to our current ecological crisis. It has shaped our major institutions—schools, government, and business—and it has shaped our everyday lives.

Fragmented thought has allowed the development of universities in which the ancient and holistic domain of philosophy has been neatly divided into separate realms of physics, chemistry, history, biology, and the like, each with its own faculty, curriculum, infrastructure, and bureaucracy—often well separated or isolated from the rest of the university. In biology alone, universities often contain separate departments of botany, zoology, and entomology, not to mention "weed science" and "plant pathology." They have further divided the pure from the applied—zoology from wildlife management, botany from range management. Our educational system has not only separated economics from ecology but also views conservation as the domain of a small subset of "applied" biologists and systematically excludes the social scientists, even with compelling evidence that most conservation problems have their roots in human activities and needs.

Fragmentation is equally pervasive in our government, where agencies charged with development are separated from those charged with conservation. Thus, for many years in this country, the Agriculture Department subsidized the draining of wetlands for agricultural purposes while the Interior Department sought to mitigate the ensuing damage. And to this day, land management agencies are typically divided into programs of wildlife, forestry, range, and so forth, each with separate leadership, funding, and direction, as if these resources existed in different landscapes.

Even our environmental "white knights"—nongovernmental organizations—have not been immune. Some fifteen years ago, Rood (1975) pointed out that some of the oldest, largest, and most respected environmental organizations in America owned stock in the very corporations and industries responsible for environmental destruction and indifference. The fragmentation of life is pervasive in our modern business world, as well, where religious devotion to the (short-term) bottom line is proudly professed at the expense of quality of life, environmental degradation, human resource development, and long-term stability. We have even invented the corporation to separate personal and business responsibilities.

Yet these institutions (university, government, nongovernmental organization, business) are no more than the embodiment of the collective actions of individuals. And it is there that the crisis of character is rooted. Only when people separate avocation from vocation, convictions from actions, or personal from corporate morality do they enable these institutions to respond the way they do. Modern humans have even developed the ability to compartmentalize blame for our current problems. Thus environmental problems are easily blamed on the businessman or the bureaucrat. The producer or harvester or regulator of a product is held guilty while the consumer remains blameless. Yet the environmentalist working weekends at his redwood deck is exploiting redwoods just as much as the tree faller and the logging company. The sportsman or birdwatcher driving half a day—using a tank of gas—to catch a big trout or see a rare bird is part of a system contributing to global warming just as is the oil company.

In summary, we have evolved into a human society in which individuals are alienated not only from nature but from the world in general. We have few ties to the land or to the places we inhabit; we often live in semi-isolation from the rest of the human community; we have neither much knowledge of our past nor much vision for the future. And while we recognize and even protest against the degradation of our environment, we appear to be quite ignorant about our own contribution to the process. We are indeed a fragmented generation.

GETTING STARTED

Given the fragmented nature of modern thinking and life, our response to threats to biodiversity has been completely true to form. To cope with these pervasive and accelerating threats we have relied almost exclusively on two strategies—both of which betray a penchant for a piecemeal approach to the problem.

The first and oldest approach is the nature preserve—in the form of a national or state park, natural area, or similar designation. "Nature" is probably the precursor word for biodiversity, and preservation of "nature" has been viewed as an activity that could be compartmentalized. The unwritten philosophy of nature preserves is that "nature" can be preserved on one side of the fence so that exploitation can continue unabated on the other. More recently, emphasis has been placed on the recovery of endangered species. This approach not only focuses on individual species, instead of the larger (eco)systems of which they are an interdependent part, but it is reactive—waiting until problems have become acute before intervening.

Both of these approaches are important parts of a biodiversity conservation strategy. But by themselves they are inadequate. The era of creation of large-scale reserves is over in this country, and the 3 percent of the land in such designation is inadequate to ensure conservation of biodiversity (Thomas and Salwasser 1989). Similarly, the species-by-species approach, by itself, is too costly, too slow, and too reactive to deal with current threats to biodiversity (Hutto et al. 1987; Scott et al. 1988).

If these programs are inadequate because they are piecemeal approaches, then we must look to a more holistic strategy if long-term solutions are to be found. Fritjof Capra has pointed out that all the major crises of our time are systemic—that is, they are interconnected and interdependent—and that they are just different facets of a single crisis which is a crisis of perception. Furthermore, he writes: "Only if we perceive the world differently will we be able to act differently. So we need a change of perception, a shift of paradigms in our thinking and in our values. We need a shift from fragmentation to wholeness, from a mechanistic view of the world to an ecological view, from domination to partnership, from expansion to conservation, from efficiency to sustainability" (Capra 1990:45). Thus to reverse the loss of biodiversity we need to think differently and act differently.

The two chapters that follow describe theory and practice that will allow us to begin integrating human activities with conservation of biodiversity. They suggest how we can begin to reintegrate both our fragmented landscape and our fragmented thinking. In Chapter 9, Gary Barrett and Patrick Bohlen describe a new holistic science, landscape ecology, that focuses on the landscape. Landscape ecology explicitly recognizes humans as part of the landscape and attempts to integrate ecological theory with practical application. They further elucidate how landscape ecology is interrelated with other fields of applied ecology that recognize a common concept: sustainability. The discipline of landscape ecology thus represents a major step away from the fragmentation of our modern university and also offers a powerful and holistic framework addressing the key issue: how to integrate human activities with conservation of biodiversity.

In Chapter 10, by Keith Hay, the practice of creating linkages through the use of greenways is described. By using greenways for preserving and linking larger expanses of parkland and preserves, we need no longer rely on parks as our islands of biodiversity. Greenways are not a panacea, of course, and not all greenways are going to be equally beneficial for conservation purposes. But something is being done here and now. Furthermore, greenways can serve as spiritual catalysts by bringing people closer to nature in their everyday lives.

The reintegration of human activity with biodiversity conservation will be slow, complex, and often contentious. It will require a rethinking and renovation of virtually all facets of our institutions and lifestyle, as well as a change in the way we manage our landscapes. The following chapters illustrate two important ways in which this process is already under way.

REFERENCES

Anon. 1990. Owl calls—the latest scuttlebutt about the northern spotted owl. *Forest Watch* 11(2):5–7.

Berger, J. 1990. You can't go home: the hidden pain of 20th century life. *Utne Reader* 39:85–87.

Berry, W. 1977. *The unsettling of America—culture and agriculture*. San Francisco: Sierra Club Books.

Bunch, W. G. 1990. Saving endangered species and the savings and loan debacle. *Endangered Species Update* 7(8–9):1–4.

Capra, F. 1982. *The turning point*. New York: Bantam Books.

————. 1990. We need a new vision. *Utne Reader* 41:45.

Dasmann, R. F. 1985. Achieving the sustainable use of species and ecosystems. *Landscape Planning* 12(3):211–219.

Edmondson, B. 1990. Making yourself at home: the baby boom generation yearns to settle down. *Utne Reader* 39:74–82.

Hutto, R. L., S. Reel, and P. B. Landres. 1987. A critical evaluation of the species approach to biological conservation. *Endangered Species Update* 4(12):1–4.

Miller, K., W. Reid, and J. McNeely. 1989. A global strategy for conserving biodiversity. *Endangered Species Update* 6(8):1–5.

Narr, K. J. 1956. Early food producing populations. Pp. 134–151 in *Man's role in changing the face of the earth* (W. L. Thomas, Jr., ed.), vol. 1. Chicago: University of Chicago Press.

Rood, W. 1975. Environment groups invest in the polluters. *Los Angeles Times*, July 20.

Sauer, C. O. 1956. The agency of man on the earth. Pp. 49–69 in *Man's role in changing the face of the earth* (W. L. Thomas, Jr., ed.), vol. 1. Chicago: University of Chicago Press.

Scott, J. M., B. Csuti, K. Smith, J. E. Estes, and S. Caicco. 1988. Beyond endangered species: an integrated conservation strategy for the preservation of biological diversity. *Endangered Species Update* 5(10):43–48.

Thomas, J. W. 1990. A conservation strategy for the spotted owl. *Forest Watch* 11(1):9–12.

Thomas, J. W., and H. Salwasser. 1989. Bringing conservation biology into a position of influence in natural resource management. *Conserv. Biol.* 3(2):123–127.

Watts, A. 1958. *Nature, man, and woman.* New York: Vintage Books.

White, L. 1967. The historical roots of our ecological crisis. *Science* 155:1203–1207.

9 Landscape Ecology

Gary W. Barrett and
Patrick J. Bohlen

LANDSCAPE ECOLOGY is an integrative paradigm that weds ecological theory with practical application (Barrett 1984, 1987). Specifically, landscape ecology considers the development and dynamics of spatial heterogeneity, the spatial and temporal interactions and exchanges across the landscape, the influences of spatial heterogeneity on biotic and abiotic processes, and the management of this spatial heterogeneity for society's benefit and survival (Risser et al. 1984). Studies of the landscape represent a new holistic, problem-solving approach to resource management. Currently there is a dire need for more long-term, cost-effective, ecological research at the ecosystem and landscape levels. Landscape ecology meets this need by providing a hierarchical perspective to help scientists understand the landscape mosaic and, especially, by affording insights into the management of landscapes damaged by both natural and human disturbances. Landscape ecology is an emerging paradigm within the area of applied ecology. The concept of sustainability is a common idea, if not the driving force, behind these new interdisciplinary fields of study. Sustainability is best achieved in landscapes characterized by natural, unsubsidized ecosystems (that is, free of man-made substances and chemicals).

The management of total landscapes will require that we understand the theory and behavior of natural ecosystems and how their processes (nutrient recycling, species regulation, biotic conservation) are both maintained and modified by natural and foreign perturbations (Barrett et al. 1990). Understanding how natural unsubsidized ecosystems and

149

landscapes function, therefore, provides insights into problem-solving approaches related to managing energy resources, maintaining genetic diversity, and reducing environmental contamination. Natural resource managers must understand the concepts of landscape sustainability—and apply them in their management practices—so that long-term solutions to resource management problems can be implemented in a cost-effective manner.

Thus, new educational and research approaches based on sustainability are urgently needed. These approaches must include new technologies (such as GIS) and encompass a hierarchical understanding of spatial scale (from forest gap dynamics to total landscapes), levels of ecological organization (from individual populations to the biosphere), and temporal scale (from short-term disturbance to global climate change). We argue here that a holistic approach, based on the concept of landscape sustainability, should underpin future resource management decisions. Landscape corridors—their definition, establishment, and conservation—will be used to illustrate this approach.

A LANDSCAPE PERSPECTIVE

To understand a landscape's patterns (such as the agricultural landscape mosaic), its elements (such as landscape corridors), and its processes (such as habitat fragmentation) requires that theory and application be integrated into a holistic research and management approach. Integrating concepts include hierarchy theory, sustainability, net energy, and systems-level regulatory mechanisms; integrating approaches include problem-solving algorithms, cost/benefit analyses, GIS monitoring, and systems analysis. These concepts and approaches must be formulated according to the questions being addressed, the temporal/spatial scale to be evaluated, and the resource management goals to be implemented. Thus, natural resource managers need to command an understanding of ecosystem, landscape, and global concepts, as well as knowing which research or management approach is most appropriate. This understanding must also include an integration of landscape theory, hierarchical approaches, historical and predicted disturbance regimes, resources to be conserved, and socioeconomic constraints.

Ecologists and natural resource managers often fail to recognize that several approaches are available for efficient and cost-effective impact assessment or resource management. But problems associated with natural or human disturbance—storm frequency, road construction, fire, acid rain—frequently are attacked only in a site-specific manner (that is, analyzed via a reductionist rather than a holistic approach). Moreover, problems are often viewed at the wrong spatial scale (that is, evaluated at the species or population level rather than at the ecosystem or landscape level). Lastly, problems are all too often assessed at the wrong temporal scale (that is, assessed in terms of immediate financial costs rather than long-term sustainable benefits). Management problems frequently arise, for example, when perturbation effects (such as pesticide application, nutrient enrichment, or acidification) are tested at one level (individual, population, community) and then used without sufficient study at another level (ecosystem, watershed, landscape). For example, one study (Barrett 1968) found that acute pesticide application which remained toxic in the environment for only a few days elicited long-term effects on litter decomposition and mammalian reproductive success. Moreover, Carson and Barrett (1980) and Levine et al. (1989) noted that municipal sludge application affected natural plant and animal species differently depending on the trophic level at which they functioned, on the type of nutrient enrichment, and on the stage of ecosystem development. A landscape approach, therefore, provides a holistic perspective which helps to ensure that efficient and cost-effective resource management occurs and that concepts such as sustainability, carrying capacity, connectivity, net energy, and minimum critical scale are encompassed in the decision-making process.

The concept of sustainability also encourages natural resource managers to view new fields of study, such as urban ecology (a heterotrophic systems approach) and agroecosystem ecology (an autotrophic systems approach), not as separate fields of study but as fields that can be coupled to provide for a sustainable landscape and a sustainable society. This approach also helps to unite ecologists (who study nature's household) with economists (who manage nature's household) to develop a common lexicon, research design, and management strategy that maintains the biotic and abiotic richness necessary for long-term sustainable ecosystems. We no longer have the time, habitat integration, or financial resources to manage wildlife resources with a single-species or single-habitat approach—only a landscape or global management perspective based on the concept of sustainability will ensure that future generations enjoy a diversified and quality environment.

LANDSCAPE CORRIDORS: TYPES AND ORIGIN

Landscape corridors (fencerows, hedgerows, stream corridors) are important structural components that frequently connect major patches (woodlots, forests, agricultural fields). Corridors are increasingly recognized as important landscape elements for such functions as affording means for animal dispersal, providing habitat for nongame species, preventing soil and wind erosion, and aiding in integrated pest management. We will define a variety of landscape corridors and illustrate how they can affect natural resource management, especially in the agricultural landscape mosaic of the Midwest. Our emphasis will be on a problem-solving approach to landscape corridors, especially as applied to resource management.

A landscape corridor is a strip of land or vegetation that differs from the extensive landscape element on either side of it. This extensive element through which the corridor runs, known as the matrix, is the most connected element present in a landscape. The landscape matrix in the midwestern United States is agricultural croplands; major corridors that run through this matrix include roads and roadside vegetation, ditches, streams, powerline cuts, and hedgerows. These corridors provide major aesthetic, economic, and ecological functions—functions that make corridors extremely important both for wildlife and agriculture and that depend to a large extent on a corridor's structure, shape, type, and geographic relation to the surrounding landscape.

Corridors can be classified into five basic types based on their origin: disturbance corridors, planted corridors, regenerated corridors, environmental resource corridors, and remnant corridors. A linear disturbance through the landscape matrix produces a "disturbance corridor." Disturbance corridors disrupt the natural, more homogeneous landscape. Nevertheless, they provide important habitat for native plant and animal species adapted to disturbance or for early successional native plant and animal species. For example, powerline corridors that cut through forested areas replace forest habitat with grasses and shrub vegetation. Forest–interior animal species seldom use such corridors to nest or reproduce. Shrub and grassland wildlife species, however, flourish in such areas. Disturbance corridors also act as barriers to the movement of some species but provide dispersal routes for small mammal species such as white-footed mice (*Peromyscus leucopus*) and eastern chipmunks (*Tamias striatus*). (See Wegner and Merriam 1979; Henderson et al. 1985.) The vegetation of transmission-line corridors must be periodically cut back, and the methods used to manage the corridor vegetation can have a significant impact, positive or negative, on the species that reside within such corridors or disperse along them. Where the vegetation is mowed, the impact on wildlife species can be reduced by staggering the mowing process (Kroodsma 1982). Management practices, for example, that increase the heterogeneity of the corridor vegetation may increase the nesting success of the mixed-habitat bird species that inhabit these corridors (Chasko and Gates 1982). Interestingly, the corridor's edge may function more as a barrier between two habitat types than as a transition zone and may also serve as a natural territorial boundary for many bird and mammal species.

"Planted corridors" do not occur naturally but are planted by humans for a variety of aesthetic, economic, and ecological functions. Some provide firewood or protection from the elements; others provide excellent habitat for predaceous insects or corridors for animal dispersal. Thousands of these corridors were planted in the Great Plains, as part of the Shelterbelt Project in the 1930s, to stop wind erosion and provide wood (Shelterbelt Project 1934).

Planted corridors are common in agricultural landscapes and can serve a variety of purposes. Grassy uncultivated corridors have been shown to help control insect pest populations within soybean agroecosystems (Kemp and Barrett 1989). For example, soybean plots divided by grassy corridors contained fewer adult potato leafhoppers (*Empoasca fabae*)—indicating the impact of grassy areas on the distribution of adults colonizing the crop. Moreover, *Nomuraea rileyi*, a fungal pathogen, infected significantly higher proportions of the green cloverworm (*Plathypena scabra*) in plots divided by grassy corridors. This fungus is the primary natural biotic control agent of this major lepidopteran larva pest in the Midwest. Thus, these planted corridors appear to play an important role in the regulation of certain insect pests within soybean agroecosystems. These findings suggest that uncultivated corridors could help regulate insect pest populations within the larger agricultural landscape mosaic.

Corridors were at one time commonly planted with Osage orange trees (*Maclura pomifera*) because these trees provided a hard, rot-- resistant wood that made excellent fenceposts. Farmers also used to plant hedgerows with multiflora rose (*Rosa multiflora*) to create a barrier to livestock movement. The common use of shelterbelts and grassed waterways to prevent wind and water erosion in agroecosystems further attests to the recognized importance of planted corridors as sound agricultural practice.

In agricultural landscapes of the midwestern United States east of the Great Plains, the most common corridors are "regenerated corridors," such as hedgerows or strips of vegetation along fences and roadsides, that regenerate from a previously disturbed area. Regenerated corridors serve a variety of functions in agricultural landscapes. They not only provide refuge for plant and animal populations but also serve as important links that allow animal movement and seed dispersal between larger habitat patches. Numerous species of insects, birds, and mammals thrive in regenerated corridors. For example, Pollard (1971) found that such corridors are important reservoirs and overwintering sites for insect predators and can act as barriers to insect movement.

Many types of animals move from hedgerow corridors into adjacent fields, including mice, birds, and insects (Pollard 1968; Pollard and Relton 1970). Although some animals may do economic damage by feeding on crops adjacent to corridors (Forman and Baudry 1984), Price (1976) found that uncultivated corridors were often the source of natural enemies that annually colonize adjacent crops and aid biotic pest control. Birds are common inhabitants of wooded regenerated corridors. Forest-edge species and those that forage in cropland often nest in wooded corridors. These bird species also help to regulate insect pest species in agricultural crops (Dambach and Good 1940). Regenerated corridors can also reduce the isolation of forest bird populations by providing conduits for movement and are, therefore, important to the preservation of biotic diversity (MacClintock et al. 1977). Red fox (*Vulpes vulpes*), white-tailed deer (*Odocoileus virginianus*), and groundhogs (*Marmota monax*) frequently use regenerated corridors. Smaller mammals, such as the wood mouse and the eastern chipmunk, have been found to suffer local extinctions in relatively isolated patches of habitat (such as forest woodlots). To survive, these small mammals depend on individuals that use regenerated corridors to reestablish the populations (Middleton and Merriam 1981; Henderson et al. 1985).

"Environmental resource corridors" are those that occur naturally where a resource is distributed in a narrow strip across the landscape (for example, a gallery forest along a stream). Vegetated stream corridors benefit the agricultural landscape because they intercept nutrients and sediments that would otherwise end up in streams causing cultural eutrophication problems (Karr and Schlosser 1978; Lowrance et al. 1984). They also reduce extreme fluctuations in stream levels. The U.S. Environmental Protection Agency has cooperated with the Soil Conservation Service and Fish and Wildlife Service to inform farmers and landowners of the benefits of these strips. The establishment of vegetative filter strips has been approved as an acceptable practice for the Conservation Reserve Program, and landowners can receive support from the federal government to establish and maintain them. These strips not only provide for water quality but, equally important, help to conserve the natural biotic diversity within the agricultural landscape mosaic.

A "remnant corridor" occurs when most of the vegetation is removed from an area but a strip of native vegetation is left uncut. These corridors are sometimes found along streams, steep land, or property lines, but they are relatively uncommon in the midwestern United States. Remnant habitats and corridors are vital as outdoor teaching laboratories depicting how "mature" ecosystems function in contrast to "young" ecosystems. For example, remnant habitats promote high species diversity, closed nutrient cycles, slow growth and yield, but increased stability. Agricultural habitats, in contrast, often promote low species diversity, open nutrient cycles, high growth and yield, but no community regulatory processes. Thus it is wise from a resource management perspective to protect these remnant habitats and corridors for a multitude of educational, economic, and ecological reasons.

Remnant corridors are probably the most important type for conserving biodiversity—especially if they are managed as natural reserves. Resource managers have traditionally focused on maximizing the amount of edge to increase diversity in the landscape (Leopold 1933; Allen 1962). But resource managers and conservation biologists have begun to recognize the need to develop a regional landscape approach (Noss 1983) that focuses on the ecological mosaic of patches with varying degrees of connectiveness and emphasizes the importance of matrix and corridors to terrestrial habitat island dynamics.

We still have much to learn about the role of patch and corridor shape as it relates to such phenomena as population dispersal, nutrient transfer rates between patches, and species extinction rates. The geometry of habitat patches is important because patches of equal size but different shapes differ in edge-to-size ratios. For example, Leopold (1933) documented that edges tend to support greater wildlife diversity than the interior of patches (a tendency commonly referred to as the "edge effect"). More recent studies of plant and nongame animal species, however, have revealed that many characteristics of edges are considered undesirable, including increased parasitism and herbivory (Laudenslayer 1986; Harris 1988; Yahner 1988). Because small, narrow patches contain more habitat edge per area than do large, square patches, edge effects are of special concern to landscape ecologists and conservation biologists. The shape and size of corridors, along with the patches they join, affect the overall success of biodiversity conservation between habitats and within regions. Thus, there is an urgent need to establish a series of long-term, integrative studies that address these ecosystem and landscape-level processes and relationships (Magnuson 1990).

Taken together, these various shapes and types of patches and corridors help create the mosaic patchwork that is common to many agricultural landscapes. Corridors may enhance biotic diversity, serve as routes for animal dispersal, and aid integrative pest management within the agricultural landscape. These functions make it essential to preserve corridors in agricultural and managed landscapes, but landscape ecologists and conservation biologists must simultaneously address the potential problems of population, community, ecosystem, and landscape interactions as well.

A STEP-BY-STEP APPROACH

Different corridors are appropriate for different landscapes. For example, a disturbance corridor (transmission line, maintenance road, or fire prevention corridor) is best suited to a forest landscape, whereas a planted corridor (windbreak, hedgerow, or intercropped corridor) enhances an agricultural landscape. We have developed a problem-solving algorithm (Barrett 1985)—an efficient, long-term, and cost-effective approach—to establish these various corridors and ensure landscape sustainability (genetic variability, reproductive fitness, and food chain dynamics). The highlights of this approach are summarized here:

1. *Problem identification*: The need for a corridor is probably related to problems associated with animal dispersal, integrated pest management, soil erosion, or the spread of disturbance. A hierarchical perspective (that is, the identification of population-level problems nested within the larger landscape-level problem) should accompany this initial step (O'Neill et al. 1986).

2. *Project coordination*: A steering committee is formed to ensure that the algorithmic process is conducted in a scientific manner. The committee's composition and size will depend on the identified problem and the temporal/spatial scale to be addressed. This group must recognize that their role is to coordinate an interdisciplinary process, not to make quick decisions regarding the establishment of a landscape corridor.

3. *Defining the universe*: The question of corridor establishment must be researched within a bounded universe. Where possible, an ecological management unit (such as a watershed) should be used rather than a political management unit (such as a county). The bounded universe must initially be viewed from a holistic, landscape-level perspective.

4. *Goal setting*: Goals are normally stated in ideal terms—for example, "to establish a landscape corridor that will maximize biotic diversity with minimal environmental degradation."

5. *Factor identification*: All major factors related to corridor establishment must be identified (costs, soil erosion or conservation, rates of animal dispersal, changes in species or landscape diversity, effects on net primary productivity within the landscape, and so forth).

6. *Information retrieval and analysis*: The coordinating committee must assure itself—and the public to be affected—that all important information (scientific publications, government reports, wildlife surveys) has been reviewed.

7. *Translation to specific objective*: With the aid of experts interacting with the coordinating committee, each factor identified in step 5 must be translated into a specific objective. (See Barrett 1985 for details.) This is a critical step in the process because all factors (aesthetics, cultural, socioeconomic, biological, and physical) must be quantified.

8. *Research design, data collection, and data analysis*: This step is similar to the scientific method: If there are no data concerning a factor identified in step 5, then data must be collected and analyzed so the factor can be translated into a specific objective (step 7). Naturally, this sequence demands a proper research design, sampling methodologies, and statistical analyses.

9. *Weighting*: Dollars, energy, and environmental quality (EQ) units are frequently used in problem-solving and impact assessment procedures (Odum 1977). All quantified factors (step 7) must be weighted—either by the coordinating committee using the Delphi method (Dalkey and Helmer 1963) or by those affected within the landscape.

10. *Devising alternative solutions*: Alternatives would likely entail proposing three or four landscape corridors (alternative sites), although only a single best corridor (solution) is to be selected. Viable alternatives (potential corridors) can be readily generated using Geographic Information Systems (GIS).

11. *Forecasting*: The weighted value for each factor (step 9) must be forecast for each alternative (that is, for each proposed landscape corridor). One corridor might be more expensive, for example, but would increase rates of large mammal dispersal and consequently increase reproductive fitness. This same alternative might also prevent soil erosion but would probably increase rates of disease transmission. Thus each factor for each alternative must be forecast in quantitative terms.

12. *Evaluating and selecting alternatives*: The best solution—that is, the corridor with the highest total EQ value—should be selected for implementation. (See Barrett 1985 regarding the statistical analysis.)

13. *Implementation and inspection*: These steps ensure that the best corridor site is established.

14. *Summary report*: The final report should be made available to all concerned (the public, state or federal agencies facing similar

landscape problems, the scientific community) and should inform them how the solution was determined and how their respective interests were treated.

By following the steps outlined here, we can achieve sound policymaking decisions that will protect the total landscape, not just preserve an individual species or community type or appease a single vested interest.

THE KEYSTONE: SUSTAINABILITY

It has become increasingly clear that more structure and habitat diversity must be introduced into the landscape mosaic. Landscape corridors help to provide these structural and functional needs. After several decades of promoting monoculture as an agricultural practice, for example, the 1990s may witness a more diversified approach—not only within individual farms (increased intercropping and lower-input sustainable agriculture) but also across the total landscape (increased habitat and landscape diversity based on a holistic, watershed perspective). The concept of sustainability must not only permeate agroecosystem ecology; it must also become a foundation for the new fields of conservation biology and landscape ecology and become fully incorporated into natural resource management. Sustainability will help to ensure that future generations will enjoy abundant resources and a quality environment. Without it, as in so many other aspects of American life, we will foolishly mortgage the future for those who follow.

REFERENCES

Allen, D. L. 1962. *Our wildlife legacy.* New York: Funk & Wagnalls.
Barrett, G. W. 1968. The effects of an acute insecticide stress on a semi-enclosed grassland ecosystem. *Ecology* 49:1019–1035.
————. 1984. Applied ecology: An integrative paradigm for the 1980s. *Environ. Conserv.* 11:319–322.
————. 1985. A problem-solving approach to resource management. *BioScience* 35:423–427.

_____. 1987. Applied ecology at Miami University: an integrative approach. *Bull. Ecol. Soc. Am.* 68:154–155.

_____. 1989. Viewpoint: a sustainable society. *BioScience* 39:754.

_____. 1990. Nature's model. *Earthwatch* 9:24–25.

Barrett, G. W., and C. A. Puchy. 1977. Environmental science: a new direction in environmental studies. *Intern. J. Environ. Studies* 10:157–160.

Barrett, G. W., N. Rodenhouse, and P. J. Bohlen. 1990. Role of sustainable agriculture in rural landscapes. Pp. 624–636 in *Sustainable agricultural systems* (C. A. Edwards, R. Lal, P. Madden, R. H. Miller, and G. House, eds.). Ankeny, Ia.: Soil and Water Conservation Society.

Callahan, J. T. 1984. Long-term ecological research. *BioScience* 34:363–367.

Carson, W. P., and G. W. Barrett. 1988. Succession in old-field plant communities: effects of contrasting types of nutrient enrichment. *Ecology* 69:984–994.

Chasko, G. G., and J. E. Gates. 1982. Avian habitat suitability along a transmission line corridor in an oak-hickory forest region. *Wildl. Monogr.* 82:1–41.

Dalkey, N., and O. Helmer. 1963. An experimental application of the Delphi method to the use of experts. *Manage. Sci.* 9:458–467.

Dambach, C. A., and E. E. Good. 1940. The effects of certain land use practices on populations of breeding birds in southwestern Ohio. *J. Wildl. Manage.* 4:63–76.

Forman, R. T. T., and J. Baudry. 1984. Hedgerows and hedgerow networks in landscape ecology. *Environ. Manage.* 8:495–510.

Harris, L. D. 1988. Edge effects and conservation of biotic diversity. *Conserv. Biol.* 2:330–332.

Henderson, M. T., G. Merriam, and J. F. Wegner. 1985. Patchy environments and species survival: chipmunks in an agricultural mosaic. *Biol. Conserv.* 31:95–105.

Karr, J. R., and I. J. Schlosser. 1978. Water resources and the land-water interface. *Science* 201:229–234.

Kemp, J. C., and G. W. Barrett. 1989. Spatial patterning: impact of uncultivated corridors on arthropod populations within soybean agroecosystems. *Ecology* 70:114–128.

Kroodsma, R. L. 1982. Bird community ecology on power line corridors in east Tennessee. *Biol. Conserv.* 23:79–94.

Laudenslayer, W. F. Jr. 1986. Summary: predicting effects of habitat patchiness and fragmentation. Pp. 331–333 in *Wildlife 2000: modeling habitat relationships of terrestrial vertebrates* (J. Verner, M. L. Morrison, and C. J. Ralph, eds.). Madison: University of Wisconsin Press.

Leopold, A. 1933. *Game management.* New York: Scribner's.

Levine, M. B., A. T. Hall, G. W. Barrett, and D. H. Taylor. 1989. Heavy metal concentrations during ten years of sludge treatment to an old-field community. *J. Environ. Qual.* 18:411–418.

Lowrance, R., R. Todd, J. Fail, Jr., O. Hendrickson, Jr., R. Leonard, and L. Asmussen. 1984. Riparian forests as nutrient filters in agricultural watersheds. *BioScience* 34:374–377.

MacClintock, L., R. F. Whitcomb, and B. L. Whitcomb. 1977. Evidence for the value of corridors and minimization of isolation in preservation of biotic diversity. *Am. Birds* 31:6–16.

Magnuson, J. J. 1990. Long-term ecological research and the invisible present. *BioScience* 40:495–501.

Middleton, J., and G. Merriam. 1981. Woodland mice in a farmland mosaic. *J. Appl. Ecol.* 18:703–710.

Noss, R. F. 1983. A regional landscape approach to maintain diversity. *BioScience* 33:700–706.

Odum, E. P. 1977. The emergence of ecology as a new integrative discipline. *Science* 195:1289–1293.

O'Neill, R. V., D. L. DeAngelis, J. B. Waide, and T.F.H. Allen. 1986. *A hierarchical concept of ecosystems.* Monogr. Pop. Biol. 23. Princeton: Princeton University Press.

Pollard, E. 1968. Hedges III: a comparison between the Carabidae of a hedge and field site and those of woodland glades. *J. Appl. Ecol.* 5:649–656.

———. 1971. Hedges VI: habitat diversity and crop pests: a study of *Brevicorgne brassicae* and its syrphid predators. *J. Appl. Ecol.* 8:751–780.

Pollard, E., and J. Relton. 1970. Hedges V: a study of small mammals in hedges and cultivated fields. *J. Appl. Ecol.* 7:549–557.

Price, P. W. 1976. Colonization of crops by arthropods: nonequilibrium communities in soybean fields. *Environ. Entomol.* 5:605–611.

Risser, P. G., J. R. Karr, and R.T.T. Forman. 1984. *Landscape ecology: directions and approaches.* Spec. Publ. no. 2. Champaign: Illinois Natural History Survey.

Shelterbelt Project. 1934. (Published statements by numerous separate authors.) *J. Forestry* 32:952–991.

Urban, D. L., R. V. O'Neill, and H. H. Shugart, Jr. 1987. Landscape ecology. *BioScience* 37:119–127.

Wegner, J. F., and G. Merriam. 1979. Movements by birds and small mammals between a wood and adjoining farmland habitats. *J. Appl. Ecol.* 16:349–357.

Yahner, R. H. 1988. Changes in wildlife communities near edges. *Conserv. Biol.* 2:333–339.

10 Greenways and Biodiversity

Keith G. Hay

GREENWAYS ARE PLAYING an important role in the conservation of biological diversity. Because greenways encompass a wide variety of linear landscape features, have been known historically by various names, and continue to be called by different terms in different regions, they often mean different things to different people. Such diversity, while being a strength for the greenways concept, makes them difficult to define.

A practical working definition for greenways is: a landscape linkage designed to connect open spaces to form protected corridors that follow natural and man-made terrain features and embrace ecological, cultural, and recreational amenities where applicable. A more formal definition is given by Charles Little, author of the outstanding book *Greenways for America*:

1. A linear open space established along either a natural corridor, such as a riverfront, stream valley, or ridgeline, or overland along a railroad right-of-way converted to recreational use, a canal, a scenic road, or other route. 2. Any natural or landscaped course for pedestrian or bicycle passage. 3. An open space connector linking parks, natural reserves, cultural features, or historic sites with each other and with populated areas. 4. Locally, certain strips or linear parks designated as a parkway or greenbelt. [Little 1990:1]

Some people assume that greenways are little more than urban recreational paths. A narrow hiking or bike path, lined with ornamental trees and mowed grass, is not a greenway. It is a path or trail. Trails are not considered greenways unless, like some sections of the Appalachian Trail, they are bounded on both sides by wide (wider than 50 feet) protected natural corridors. Trails are in fact components of many greenways, especially in urban areas where well-developed multiple-use trails are common. In rural areas, greenways may have unimproved trails or none at all, and access to some is limited to waterways.

The idea of linking natural open spaces in the United States for beautification, recreation, and environmental protection goes back at least to the turn of the century. Many early landscape architects in America had been trained in Britain or Europe or had traveled extensively abroad and were influenced by Old World landscapes, gardens, and parks. Frederick Law Olmstead (1822–1903), unquestionably one of our greatest landscape architects and park designers, is considered by many to be the father of greenways, although he never called them that. His parks and parkways remain very much a part of our modern landscape from Central Park in New York City to the "parkway" on the Berkeley campus of the University of California. The term "greenway" was first used by the distinguished planner and author William H. Whyte in his book *Securing Open Space for Urban America* (1959).

The modern greenway concept appeared in the late 1960s and 1970s when federal funding for open space acquisition began to dry up. Local civic and environmental leaders were forced to look at areas close to home that could be acquired with little or no money through such innovative means as cooperative agreements, easements, local tax incentives for donations, ordinances, rights-of-way abandonments, and liability relief for landowners. These areas included overlooked streamsides, floodplains, industrial sites, abandoned rail rights-of-way, and utility corridors.

Saving open spaces by local initiative was certainly not a radical concept, but the idea of linking them into extensive natural corridors across the countryside is new. This grass-roots greenways movement was confirmed by the President's Commission on Americans Outdoors and became the centerpiece of its 1987 recommendations. That same year, The Conservation Fund, a national nonprofit land and water conservation organization based in Arlington, Virginia, designed a comprehensive program to advance the greenways concept and created the American Greenways program. This was followed by an intensive three-year study to examine successful greenway projects, their history, and their conceptual and physical definitions across the nation. Who was building greenways and why? What kinds were being built? How were they going about it? What problems were encountered and how were they being resolved? Over ninety projects in thirty states were examined. The results of this research have been condensed into a comprehensive book: *Greenways for America* (Little 1990).

It soon became evident that greenway projects could be categorized into several different types: rivers through a city; paths and trails; ecological corridors; scenic drives and historic routes; and greenway networks. While we will be considering all types here, our primary focus will be on greenways designed to protect their ecological values.

THE VIRTUES OF GREENWAYS

The mobility of wildlife species and their ranging requirements for food, cover, and mates is a fundamental tenet of wildlife ecology. Aldo Leopold (1933:125), nearly sixty years ago, eloquently addressed this point when he said: "The essential difference between a deer and a man is that man builds farms, factories, and cities to provide himself with the elements of an habitable range, whereas deer must accept the random assortment laid down by nature and modified by human action, or move elsewhere. In both cases that endless competition which we call society consists essentially in a struggle for the best assortment of places to feed, hide, rest, sleep, play, and breed." He discussed at length the "cruising radius" of species on a daily, seasonal, and annual basis; the need for protected "streets" between habitats; and the fact that natural mobility tends to prevent inbreeding. Elton (1930) had surmised three years earlier that disharmony within an environment stimulates animal movement and constantly disturbed habitats could result in mobility permanently stimulated.

Today, six decades later, new disciplines (landscape ecology, conservation biology, and restoration ecology) are addressing Leopold's "struggle for the best assortment of places" and how to reach them in an increasingly stimulated and fragmented landscape. (See Noss and Harris 1986; Soulé and Wilcox 1980.) The resulting loss of biological diversity at the ecosystem, species, and gene levels has now become a global concern and a domestic land management goal. Indeed, the biodiversity crisis is predicted to reach a crescendo in the first half of the twenty-first century (Soulé 1987).

Global warming—and its ecological and genetic significance for plants and animals in the next twenty-five to fifty years—is of grave concern to scientists studying conservation biology. Significant changes are anticipated in precipitation patterns, rising ocean levels, altered soil chemistry, wind and current circulation, and nonuniform temperature shifts around the world. In September 1989, several hundred scientists were convened by the World Wildlife Fund at the National Zoo in Washington, D.C., to address these implications for species survival. They concluded that the resulting changes in habitats would cause the ranges of many species to shift outside the boundaries of the parks and refuges now established to protect them. They recommended the creation of corridors—greenways—between preserves and their gene pool libraries to provide essential travel routes, thus preserving biotic diversity.

Greenways, although no panacea for the crisis, are being used as useful tools to conserve biological diversity. Linkage is the central theme and goal of the greenway concept—to reconnect and preserve natural land and water habitats, thus reversing the biologically destructive effects of landscape fragmentation that inevitably result from urbanization. Fragmentation is one of the principal causes of loss of species diversity and may lead to extinction for numerous plants and animals known to professionals as "area-sensitive" species (Harris 1985).

Riparian habitats are known to be some of our most biologically productive and diversified faunal and floral systems. The lands bordering creeks, streams, and rivers and their floodplains constitute the backbone of most greenways today and provide the potential for millions of miles of protected greenway corridors in the future. When combined with other lineal landscape features, such as coastal strips, ridgelines, segments of utility rights-of-way, parks, refuges, lakes, meadows, ravines, wetlands, and forests, greenway linkages clearly become a powerful tool for habitat connectivity. Harris (1985) points to the long-term paleontological record as evidence that the corridor approach works by promoting genetic interchange in two directions and serves to maintain full and balanced plant and animal communities in both interconnected areas. Examples of "ecological" greenways where public access is minimized or nonexistent are the Sudbury

Valley Bay Circuit Greenway in Massachusetts, the Arkansas River Greenway in Colorado, the Stony Brook Greenway in New Jersey, the Oconee Greenway in Georgia, and the Willamette River Greenway in Oregon.

Today the growing interest in wildlife conservation (including biodiversity) and lineal outdoor recreation must be channeled to the mutual benefit of both. They are not mutually exclusive. Greenways in urban regions are, of necessity, heavily oriented toward public access and recreational use. With few exceptions, however, most still retain uninterrupted ribbons of natural vegetation that extend into the countryside and retain a substantial degree of corridor integrity for a great number of plant and animal species (Adams and Dove 1989).

Abraham Lincoln once remarked that "public sentiment is everything. With public sentiment nothing can fail, but without it, nothing can succeed." Such advice is eminently appropriate to those involved in building greenways and biological diversity. Trail systems can be carefully sited and designed to ensure minimal impact on the ecology of the greenway corridor. The building of greenway systems takes time, money, patience, technical knowledge, and partnerships. Without dedicated people, the greenway corridors that protect the habitat linkages so essential for biodiversity would not happen. Thus it is only prudent that greenways provide, where possible, certain human benefits such as limited public access while minimizing any impact on the use of the corridor as a natural habitat and thoroughfare.

There are many places where functioning recreation and wildlife corridor systems are providing passageways for a large number of free-ranging mammals including moose, elk, deer, bear, javelina, coyote, fox, beaver, and numerous reptiles, amphibians, and birds. Some of our major cities are good examples: Anchorage, Alaska, has a greenbelt that stretches 12 miles around and through the city to the Chugach Mountains; in Oregon, Portland's 5,000-acre Forest Park is connected to the Coast Range by a natural corridor via the Tualatin Mountains; in Arizona, Tucson's Rillito River and its floodways and washes have been turned into a network of greenways that bring a large variety of desert species in and out of the city; and in our nation's capital, white-tailed deer, beaver, raccoons, and foxes freely travel between the historic Rock Creek Park and the wilds of West Virginia via the C & O Canal and the Potomac River.

MINIMIZING THE PROBLEMS

Greenways can be designed and maintained to protect and often enhance their ecological systems and the complex relationships among the indigenous species. Fundamental to this effort are regional management and design differences (urban/suburban versus rural; wet versus arid; lowlands versus mountains).

The ecological integrity of a greenway is protected by recognizing and eliminating, to the extent possible, its potential disadvantages: the possible spread of fire, contagious diseases, parasites, pests, exotic species, hybridization of related taxa, and increased exposure of animals to predation, domestic animals, and poachers. Such threats can be minimized by ensuring that greenways link only previously connected habitats—not naturally isolated ones. This axiom is especially important when using segments of utility rights-of-way that may cross such areas or expose a natural corridor to the problems cited above.

Another greenway problem associated with biodiversity is the potential impact of recreational development. How will the influx of people, horses, dogs, bicycles, all-terrain vehicles, camping, motorized boats, hunters, and so forth affect the natural system and its inhabitants? Here again, professional design and careful management can minimize or eliminate such negative multiple-use interactions. Duever (1988) has specifically addressed human ecological considerations in trail and greenway planning and offers a comprehensive list of topics to be considered:

- *What are the region's important natural features?* Where are they, and what are their significance, their sensitivity, their management problems? Are important tracts adequately connected to similar habitats? Are there sufficient buffers?
- *What are the topography, geology, and soils like?* What trail and maintenance problems are envisioned? Will the terrain encourage people to stay on trails? Will sinkholes, bogs, ravines, and rocky outcrops present problems?
- *What are the hydrological features?* What lakes, rivers, and springs does the region have? How changeable are the stream channels and

how deep? How frequent are flash floods and what is their impact on trail/greenway design features?

• *How does fire function in the ecosystem?* Are habitats maintained by fire? How severe and frequent are fires? Is controlled burning done? How would greenway designs mesh with fire management plans?

• *What are the vegetation dynamics?* What plant communities are present and what condition are they in? Which species are important? How do dominant and keystone species disperse and become established? What are the management needs of rare and endangered species?

• *What are the pertinent wildlife ecology issues?* Are there rare and endangered species? What are the management needs for biodiversity? Where are the breeding sites, feeding areas, and special habitats? Which animals are sensitive to human presence? What travel routes are apparent? What is the effect of design on travel corridors and the spread of pests and disease? Are animals likely to endanger trail users or vice versa?

• *What are the potential pollution problems?* Are pesticides or herbicides used in the vicinity? What kinds are used, how, and when? How serious is the problem? What are the water quality implications?

There are undoubtedly other factors that should be examined, but these considerations present an excellent base for minimizing problems associated with recreational use of greenways while maximizing ecological assets, user safety, and enjoyment. Each greenway design must be addressed differently, depending on the geographical, biological, and recreational parameters. Although greenway corridors are not problem-free, their advantages far outweigh their disadvantages. (See Soulé et al. 1988.)

THE BUILDING PROCESS

The process of developing greenways also varies with the type of greenway, its ecological potential and geographic locale, and the partnerships and sociopolitical environments involved. Making greenways happen always means developing tailor-made game plans and forging partnerships at all levels of government and with businesses, civic organizations, planners, landscape architects, developers, landowners, scientists, and, above all, interested citizens. The greenway process is still largely a grass-roots, citizen-led, community effort. But Maryland's efforts to develop a statewide greenway network and the federal government's technical assistance program developed by the National Park Service illustrate meaningful roles and opportunities for government.

Ideally, then, the process of creating greenways requires cooperative action and responsibility from the bottom up as well as from the top down. Because local greenway projects often stop at municipal or county boundaries, a statewide greenway planning approach is necessary. To address interstate or international greenway systems, or those on federal lands, a national strategy and federal responsibility are required. Let's briefly examine the strategy for creating greenways starting at the grass-roots level and working up.

The Role of Citizens and Communities. At the local level, successful greenways are being created in a variety of ways. Some are produced by the initiative of one or more citizens who design them according to local resources and the priorities of community leaders and residents. Others are entirely the product of local municipal governments. Land trusts—private, nonprofit, land conservation groups of varying size—are another successful approach. These citizen organizations are among the fastest-growing open space groups in the nation. There are some 800 active trusts with a collective membership of over 700,000. Their national organization in Washington, D.C., the Land Trust Alliance, plays a key role in the greenway movement.

One of the most successful organizational approaches has been the quasi-governmental greenway commission. These public/private commissions (or foundations), publicly authorized, are composed of a board of directors representing key citizen interests, corporations, utilities, professional associations, and local government representatives. Like land trusts, they can move quickly and decisively on land acquisitions. Moreover, they possess authority and political clout, are not hampered by a public agency's jurisdictional boundaries or narrow mandate, are largely free from political pressures, and have the capacity to raise money and receive donations, grants, and membership fees. To ensure that biodiversity considerations are an integral part of the planning process, conservation biologists should be consulted or be invited to join the greenway advisory group.

With an organizational structure in place, sponsors can then proceed with the basic elements of the greenway-building process:

1. Defining the role and function of the organizing body and setting general goals and specific objectives.
2. Conducting a ground-level reconnaissance of the proposed greenway—including an inventory and evaluation of its natural, historical, cultural, and recreational assets.
3. Developing overlays of the survey information on a working map.
4. Researching ecological, geographic, historic, and ethnographic studies; corridor landownership, rights-of-way, zoning classifications, and open space set-asides; and regulatory and administrative alternatives for protecting the greenway corridor.
5. Analyzing gaps between open spaces and other greenways (potential or existing) and establishing priorities for linkage acquisitions or easements.
6. Conducting public attitude surveys and preparing a public involvement plan that will develop a greenway constituency and a consensus for action.
7. Identifying and analyzing major public issues associated with the project—including political support.
8. Communicating the concept to the community (and appropriate government bodies) through the preparation of the following:

   · A professional, diagrammatic greenway landscape design plan
   · A well-written, colorful, and compelling "vision document" setting forth the community benefits to be derived from the greenway—including recreation and economic advantages (increased business income, higher real estate values, new jobs, and new investments)—and the ecological benefits of protecting species survival and biodiversity
   · A slide or video presentation
   · Press conferences, speeches, public walks, and public hearings

9. Initiating a funding campaign to complete a highly visible demonstration segment of the greenway. Most successful projects are built segment by segment, completing each before moving on. The completed segment offers immediate public relations benefits that help build support for the entire greenway.

10. Developing a long-range implementation strategy that sets forth detailed recommendations and priorities for construction, management, and maintenance of the greenway and plans for future connections with other greenways to form a regional or statewide network.

To be effective, all elements of this process must be carried out in a manner that engenders public enthusiasm and support.

The Role of the State. In the process of greenway planning, the state's role complements the work of local communities by providing monitoring and structure for a statewide system. The natural linearity of rivers and their tributaries, plus the overland connecting links that use ridges and rights-of-way, form the foundation for such a system. As at the community level, a quasi-governmental commission or foundation, legislatively authorized, is the organizational approach generally taken by progressive states. An enthusiastic governor, reinforced with strong support from the legislature, is of the utmost importance in funding and carrying forward the work of a successful state greenway commission. A real estate transfer tax—many states now have one—is an invaluable asset for funding greenway programs.

One of the first statewide greenway efforts is currently under development by the state of Maryland. Pushed by its governor, it is financed by a real estate transfer tax that provides a grants-in-aid program encouraging local governments and nonprofit groups to initiate greenway projects under the state's comprehensive plan. The inventorying and mapping methodology for this ambitious program is being developed by The Conservation Fund. In addition to developing internal greenway programs, state officials should also anticipate extending their greenways into adjoining states by means of memorandums of agreement.

Because greenways usually follow natural land and water corridors, state fish, wildlife, and parks departments should establish a greenway or wildlife corridor program to evaluate properties for their biodiversity and linkage potential with adjacent public and private open spaces. All land acquisition programs should be conducted with this end in mind. For example, a high priority should be given to the IQ (isolation quotient) of a proposed wildlife habitat or park purchase. Each parcel should be evaluated for its natural linkage potential with other suitable open spaces prior to purchase. This process also provides the agency with new opportunities for biodiversity conservation, nongame wildlife programs, an expanded role in community relationships, a new and larger public constituency, and very possibly a new source of funding.

The Role of Federal Agencies. Federal land management agencies have a tremendous potential for developing greenway corridor systems for biodiversity by linking fragmented habitats on their lands, between agency lands, and with adjoining public and private open spaces where appropriate. To date, no national strategy exists to encourage greenway corridors that cross agency boundaries. Few agencies have demonstrated an interest in greenways or developed linkage-analysis teams and programs. The only exception is the National Park Service's Recreation Resources Assistance Division and its Rivers and Trails Conservation Assistance Program. This program provides technical assistance to state and local governments—and even to nonprofit groups and municipalities—for assessments of river and trail corridors and open space. The Conservation Fund is cooperating with this division in the development of two important resource studies entitled *The Economic Impacts of Protecting Rivers, Trails, and Greenway Corridors* (Murray 1990) and *The Ecology of Greenways* (still in progress).

The U.S. Fish and Wildlife Service recently signed a memorandum of understanding with The Conservation Fund to implement its estuary programs in the Northeast and to develop a national greenways program that would emphasize the protection of wetlands and other wildlife habitat. Although the Endangered Species Habitat Protection Program, the North American Waterfowl Plan, the Emergency Wetland Resources Act, the Fish and Wildlife Coordination Act, and the Migratory Bird Conservation Act all have some applicability to corridor protection, development, and biodiversity, their specific contributions have yet to be defined.

The Bureau of Land Management, controlling more land surface in the West than any other public landowner, should initiate a pilot program to examine possible connections between their extensive fragmented lands and other federal, state, and local open space habitats. They not only have the authority for land exchanges, but literally millions of acres available for such programs. The Department of Defense, now reevaluating its enormous landholdings for a kinder and gentler future, also has opportunities for landscape linkages and should be considering the greenway concept and its biological benefits. The National Oceanographic and Atmospheric Administration (NOAA) has a Coastal Zone Management Program with a budget of some $30 million for state block grants with low matching requirements. Greenway planning on tributaries connecting with coastal waters and marine estuaries would be eligible for funding under this program.

Like other land managing agencies, the USDA Forest Service can play an important role by modifying its land management policies to encourage corridor development and biodiversity within national forest boundaries and with adjacent public and private properties. Some 80 percent of national park boundaries, for example, adjoin national forests. Their current emphasis on conservation biology and urban forestry programs is already proving to be a distinct asset for greenway planners, and the agency's responsibility for part of the national system of scenic and historic trails gives it a pivotal role in long-distance greenways.

The Environmental Protection Agency's mapping program for sensitive and threatened wetland habitat also holds potential for greenways. This information assists EPA in determining which areas to avoid when approving Section 404 permit applications. The EPA is using a compilation of Fish and Wildlife Service data, U.S. Geological Survey satellite imaging, and NOAA resource data on a GIS system. As part of its regulatory permitting process the EPA could offer this information to county planning commissions as an aid to restoring connections between sensitive wetlands and to avoid destruction of habitat corridors. Further, considerable money is spent by EPA on state programs for nonpoint pollution control. Greenways with riparian vegetative filter strips are very effective in preventing nonpoint pollution and improving water quality.

All these federal agencies should play a lead role in developing a national greenway strategy. Perhaps an interagency instrument of commitment is in order to define cooperative goals and initiate pilot programs.

THE GREENWAY POTENTIAL

Building greenways is a time-consuming, sensitive, diplomatic process of creative problem solving. From beginning to end, public support is crucial to its success. All of the theory, research, and learned treatises on landscape ecology, biodiversity, and associated sciences mean little unless such knowledge can be put to use on the ground. Creating and protecting greenway corridors can become the basis for achieving many of these research goals and scientific missions.

The contributions of this or that greenway to the conservation of biodiversity will naturally vary. The greenway corridor concept, however, holds more potential for protecting plant and animal communities and their genetic richness—while improving the quality of our everyday life—than any other current land management program.

REFERENCES

Adams, L. W., and L. E. Dove. 1989. *Wildlife reserves and corridors in the urban environment*. Columbia, Md.: National Institute for Urban Wildlife.

Duever, L. C. 1988. Ecological considerations in trail and greenway planning. In *Proceedings of the National Trails Symposium*. Gainesville, Fla.: KBU Engineering and Applied Sciences.

Elton, C. 1930. *Animal ecology and evolution*. London: Humphrey Milford.

Forman, R. 1987. The ethics of isolation, the spread of disturbance, and landscape ecology. In *Landscape heterogeneity and disturbance*. (Monica Goigel Turner, ed.). New York: Springer-Verlag.

Harris, L. D. 1985. *Conservation corridors—a highway system for wildlife*. Winter Park: Florida Conservation Foundation.

Leopold, A. 1933. *Game management*. New York: Scribner's.

Little, C. E. 1990. *Greenways for America*. Baltimore: Johns Hopkins University Press.

Murray, R. 1990. *Economic impacts of protecting rivers, trails, and greenway corridors*. San Francisco: National Park Service, Western Region.

Noss, R. F., and L. D. Harris. 1986. Nodes, networks, and MUMs: preserving diversity at all scales. *Environ. Management* 10(3):299–309.

Soulé, M.E. 1987. History of the Society for Conservation Biology: how and why we got here. *Conserv. Biol.* 1(1):4–5.

Soulé, M. E., and B. A. Wilcox. 1980. *Conservation biology: its scope and challenge.* Sunderland, Mass.: Sinauer.

Soulé, M. E., et al. 1988. Reconstructed dynamics of rapid extinctions of chaparral—requiring birds in urban habitat islands. *Conserv. Biol.* 2:75–92.

Thomas, J. W., and H. Salwasser. 1989. Bringing conservation biology into a position of influence in natural resource management. *Conserv. Biol.* 3(2):123–127.

Whyte, W. H. 1959. *Securing open space for urban America: conservation easements.* Washington: Urban Land Institute.

Discussion

The following discussion reflects the audience's questions concerning the conflicts between human activities—such as development and population growth—and biodiversity conservation. Can the two be effectively integrated?

JIM PISSOT (NATIONAL AUDUBON SOCIETY): In listening to the panelists today, it seems to me that wildlife needs less management and humans need more. Population growth, industrial urban growth, and agricultural systems—those are the real problems. I have two related questions—one for Dr. Scott and one for Dr. Salwasser.

In recent hearings on biodiversity in Washington, the directors of the BLM and Forest Service stated that no new legislation is necessary to ensure management for biodiversity conservation; that sufficient legislation exists; and that in fact these agencies are carrying out the letter of the law. I would submit that a gap exists between the statement of policy and its implementation on both Forest Service and BLM lands. For example, Audubon Society maps show that timber harvesting, clearcuts, and a variety of so-called manipulations of the land in national forests are designed, in fact, to fragment those very large stands that are remnants of the ancient forest in the Pacific Northwest. The American public and Congress seem to assume that public lands are managed for biodiversity. And yet on your map, Dr. Scott, you appear to have lumped these public lands in the same category as the City of Boise and the potato fields of southern Idaho. My question then, Dr. Scott, is this: Do you really consider these national forest and BLM lands to be managed for biodiversity? And, Dr. Salwasser, when will we see the kind of differences on national forest lands that we have been promised in terms of protecting biodiversity?

176

MICHAEL SCOTT: We do not consider Boise and the rest of urban southern Idaho in the same category with the national forest. There are a lot of opportunities for creative management that would increase the protection of biodiversity in Idaho, but they would have to be codified and would not involve locking up the remainder of federal lands in Idaho. For example, suppose we were to find an area of vegetation type that is completely unprotected in Idaho. And suppose that in looking at our landownership map, we were to discover that 80 percent of that vegetation type is on Forest Service land. We would first go to The Nature Conservancy and ask them to identify the best elements of that particular vegetation type. Then we would go to the Forest Service with that information and make a case for some sort of management scheme to ensure the long-term viability of that area. I use the term "management scheme" rather than "preserve" because I think that is what we are going to be looking at in the long term.

HAL SALWASSER: To take a stab at answering your question, Jim, the national forests are managed under a long history of laws that represent the American people's will for what they want their national forests to do. And they call it multiple use—which means there are going to be areas that are managed intensively to provide resources to people and areas that are managed to protect certain things. It's a struggle to figure out how to balance those two goals, especially when they come into conflict, as they do all too often in the Pacific Northwest right now.

Based on changes in people's expectations, new knowledge, and new technologies, we are trying to make adjustments in our land management plans; we are trying to bring on new technologies. I am not going to tell you that clearcuts in the Olympic National Forest are good for biodiversity. They are not. They are, however, pretty good for meeting people's short-term needs for raw wood fiber materials.

In terms of progress, the Forest Service has done a lot to strengthen the threatened and endangered species program and has taken major initiatives in fisheries and the whole wildlife sector. All I can say is we are working as hard as we can to bring better balance to the overall application of a multiple-use mission. It does not mean we are going to protect biological diversity on every site because that is not the congressional mission of the agency.

HOPIE STEVENS: Where I live in Montana, there's a lot of concern among timber communities about jobs. I think we have an obligation to find alternative means of living for people in areas where timber production has been replaced by preservation or conservation measures. I wonder if Dr. Salwasser has anything to say on this subject?

HAL SALWASSER: It's important to approach enterprises like forestry or range management so that you have long-term sustainability of the business. It doesn't mean you have to stop logging; it means that you carry out your harvesting practices and your reforestation practices in a way that can sustain a productive forest. I think there are many examples across the United States where that's been done.

The notion of managing the human population has popped up in the discussion a number of times today. I think it's critically important that as we think about taking people out of the lifestyle they're in, without physically moving them out of their environment, we must give careful thought to the potential adverse impacts of the alternative lifestyles we offer. Anything you do to change the relationship of people with the land is not a simple task—and will result in changes in peoples' lifestyles, as well.

UNIDENTIFIED SPEAKER: All this talk of managing the land base leaves me a bit concerned that we are losing sight of the tried and true methods such as captive propagation and restoration. Restoration is not a land-based question; it is a matter of public opinion, of adequate funds, and of sufficient manpower. I hope we don't lose sight of that in trying to create the butterfly net of biodiversity conservation that attempts to catch all species. Some species simply require additional attention.

MICHAEL SCOTT: There is no question that captive propagation is an extremely valuable tool in terms of the restoration process of biological diversity. For some species, like the California condor and the black-footed ferret, that is our only hope, our last chance. However, my colleagues and I have recently completed a survey of all captive propagation and restoration programs for game species and threatened and endangered species throughout North America, Hawaii, New Zealand, and Australia. We found that, with the threatened and endangered species, only 15 percent of the programs for captive propagation and captive rearing for release or translocation were successful in restoring the self-sustained wild population. Additionally, when captive-reared animals are used in a translocation, your chances of having a successful program are reduced by 50 percent. We can't wait to fall back on this expensive, last-ditch approach; we must act now before the situation reaches crisis proportions.

EDITH THOMPSON (MARYLAND DEPARTMENT OF NATURAL RESOURCES): I have a comment that pertains to cooperation between constituencies in the public and private sectors. Maryland is a rapidly developing state—which, of course, is a big problem for wildlife and natural habitat. The people who make all the decisions about what happens to this habitat are the county land planners, the developers, the engineers, and the guy who drives the bulldozer. By actually going on site, I've found that people don't know anything—and if they do, they don't think it's important. So we've started a program called Natural Design Development where we bring all these people together through workshops designed to heighten their understanding and cooperation.

FELICE PACE: We have a similar program that we run with the California Department of Fish and Game where we teach gold miners how to mine while, at the same time, enhancing fish habitats. There's a lot of good that can come from these programs. But we shouldn't assume that the guy who drives the bulldozer doesn't care. There's a tremendous amount of support for wildlife out there and it's up to us to show people how they can protect it. Education is a key to biodiversity conservation.

MICHAEL SCOTT: I think if you're looking for a model, a group that's been able to bring together successfully diverse interests *and* to accomplish big things, a good example is The Nature Conservancy.

KEITH HAY: Just last week, a new statewide greenway initiative for the state of Maryland was announced. Their greenway open space program is one of the first in the nation. It's presently a system that's being driven from the state level down. I worry about that and I'm sure they recognize, too, that grass-roots involvement is a necessity for the successful development of a state greenway network.

EDITH THOMPSON: It's hard to get beyond our territoriality. That's one of our biggest problems, I think.

RAY OLSON (UNIVERSITY OF MAINE): In Maine, we have a somewhat different problem. We are the most forested state in the nation, but we have the least amount of public lands and the largest percentage of out-of-state industrial ownership. To be blunt, biodiversity is not an issue in Maine. Does anyone on the panel have any suggestions on how we *can* make this an issue?

HAL SALWASSER: Basically, you have two options. One is incentives and the other is regulation. I recommend that first you explore every possibility to create incentives and recognize that in doing so you probably won't get everything you want. You'll also have to have a forest practices act that has standards to achieve whatever goals you can accomplish politically. There is nothing magical about that advice.

About the Contributors

KEVIN ATKINS is director of environmental sciences at Henigar and Ray Engineering Associates, a consulting firm of architects, engineers, ecologists, planners, and surveyors. Atkins has been an environmental and wildlife consultant since 1973. He specializes in conducting environmental impact assessments of threatened and endangered floral and faunal species as related to habitat loss. Although his focus is primarily in Florida and the southeastern United States, Atkins has extensive international consulting experience and the firm has recently opened an office in Seattle.

GARY W. BARRETT is Distinguished Professor of Ecology in the Department of Zoology at Miami University in Oxford, Ohio. He is the founder and codirector of the Ecology Research Center and founder of the Institute of Environmental Sciences at Miami University. Barrett has also been president (1988–1990) of the U.S. section of the International Association for Landscape Ecology. He has published widely in numerous professional journals and is actively involved in several professional societies.

PATRICK J. BOHLEN is a graduate research fellow in the Department of Entomology at Ohio State University in Columbus. His research has focused on the effect of habitat and landscape structure upon insect population dynamics and dispersal in agroecosystems and agrolandscapes. Bohlen is currently working on his doctorate in soil ecology at Ohio State University.

BENNETT A. BROWN is director of The Nature Conservancy's Rocky Mountain Heritage Task Force in Lakewood, Colorado. He coordinates the Conservancy's science support efforts and serves as liaison with the network of state Natural Heritage programs and Conservation

Data Centers in the western United States. Brown is also involved in strategic planning for several of the Conservancy's bioreserve projects in the West. Included among these are Gray Ranch in New Mexico—the largest private conservation acquisition in the United States—and the Conservancy's Yellowstone Lands project in the northern Rockies.

STEVEN CAICCO is a consultant in conservation biology. He was formerly a plant ecologist with the University of Idaho and, prior to that, a research scientist for the Idaho Natural Heritage program. Caicco is the recipient of the American Horticultural Society's Wildflower Rediscovery Award (1984).

DOUGLAS H. CHADWICK, a Montana biologist and writer, is the author of *The Kingdom*, a book on American wildlife, and *A Beast the Color of Winter*, a book resulting from seven years of field research on mountain goats in the Rockies. Chadwick writes frequently for *Defenders Magazine* and *National Geographic*.

ALLEN COOPERRIDER, formerly a wildlife biologist with the U.S. Bureau of Land Management in Denver, Colorado, is currently a consultant in conservation biology with Big River Associates in Ukiah, California. He has twenty years of experience integrating wildlife conservation with human uses of wildlands, particularly timber harvesting and livestock grazing.

BLAIR CSUTI is an adjunct associate professor at the University of Idaho in Moscow. Formerly with The Nature Conservancy (1979–1986) and the Center for Conservation Biology at Stanford University (1986–1988), Csuti is currently director of the Oregon Species Richness/Gap Analysis Program in Portland. He has published in numerous professional journals.

LARRY D. HARRIS is a professor in the Department of Wildlife and Range Sciences at the University of Florida in Gainesville, where he specializes in forest ecosystems and biodiversity conservation. He is the author of more than fifty scientific publications and sits on several national and international boards and advisory panels. Harris's book *The Fragmented Forest* (1984) won national and international awards for excellence.

KEITH G. HAY is director of the American Greenways Program for The Conservation Fund in Arlington, Virginia. He has worked for a variety of state and federal agencies, including the Colorado Department of Fish and Game, the U.S. Bureau of Outdoor Recreation, and the U.S. Fish and Wildlife Service. Before coming to The Conservation Fund, Hay was conservation director of the American Petroleum Institute. He is actively involved in the National Institute of Urban Wildlife and lectures around the country on greenways.

REED F. NOSS is a consultant specializing in biodiversity conservation and lives in Corvallis, Oregon. Noss has worked for the Ohio Natural Heritage Program, the Florida Natural Areas Inventory, and the U.S. Environmental Protection Agency. The recipient of several research awards and commendations, Noss was recently awarded the National Wildlife Federation's prestigious Environmental Publication Award. He was a participant in the biodiversity policy dialogue sponsored by The Keystone Center and is on the editorial board of *The Natural Areas Journal*.

FELICE PACE is director of the Klamath Forest Alliance, a grass-roots coalition of environmental organizations in Etna, California. Pace has spent ten years as a citizen activist working on forest management issues in the Klamath National Forest. He is currently vice-chair of the California Ancient Forest Alliance and vice-president and conservation chair of the Marble Mountain Audubon Society. A self-trained naturalist, Pace has fifteen years' experience operating adventure-based outdoor education programs.

HAL SALWASSER is director of the USDA Forest Service's New Perspectives program in Washington, D.C. He has served as coordinator of the Forest Service's Biodiversity Strategy and as deputy director of its Wildlife and Fisheries Section. Salwasser has published widely on ecology and ecosystem management and was recently elected to the governors of the Society for Conservation Biology.

J. MICHAEL SCOTT is a research biologist with the Cooperative Fish and Wildlife Research Unit of the U.S. Fish and Wildlife Service at the University of Idaho in Moscow. He has had rich and varied experience, first as the station leader at the Mauna Loa Field Station Volcano in Hawaii and later as director of the California Condor Research Center. Presently, Scott is unit leader for the Idaho Gap Analysis Program. He is the recipient of the Wildlife Society's Best Monograph Award (1987), the University of Idaho's College of Forestry Outstanding Researcher Award (1990), and Renew America's National Environmental Achievement Award (1990).

MICHAEL E. SOULÉ is a professor and chair of the Board on Environmental Studies at the University of California in Santa Cruz. He is the founder and first president of the Society for Conservation Biology, which produces the quarterly journal *Conservation Biology*. Soulé is also a consultant to the National Park Service for its program in conservation biology.

DONALD M. WALLER is a professor in the Department of Botany at the University of Wisconsin in Madison. He has conducted field research on the demography and population genetics of rare plant populations, the evolutionary dynamics of plant breeding systems, and the impact of landscape management on plant/herbivore interactions. He has published in numerous journals and has helped organize a new graduate program in conservation biology at the University of Wisconsin.

About the Editor

WENDY E. HUDSON is communications coordinator for the Western Regional Office of Defenders of Wildlife, where she is active in promoting the organization's biodiversity and "watchable wildlife" programs. Hudson has worked on environmental subjects for ten years and has produced several conference proceedings on related issues.

Index

ALSO AVAILABLE FROM ISLAND PRESS

Ancient Forests of the Pacific Northwest
By Elliott A. Norse

Balancing on the Brink of Extinction: The Endangered Species Act and Lessons for the Future
Edited by Kathryn A. Kohm

Better Trout Habitat: A Guide to Stream Restoration and Management
By Christopher J. Hunter

Beyond 40 Percent: Record-Setting Recycling and Composting Programs
The Institute for Local Self-Reliance

The Challenge of Global Warming
Edited by Dean Edwin Abrahamson

Coastal Alert: Ecosystems, Energy, and Offshore Oil Drilling
By Dwight Holing

The Complete Guide to Environmental Careers
The CEIP Fund

Economics of Protected Areas
By John A. Dixon and Paul B. Sherman

Environmental Agenda for the Future
Edited by Robert Cahn

Environmental Disputes: Community Involvement in Conflict Resolution
By James E. Crowfoot and Julia M. Wondolleck

Forests and Forestry in China: Changing Patterns of Resource Development
By S. D. Richardson

The Global Citizen
By Donella Meadows

Hazardous Waste from Small Quantity Generators
By Seymour I. Schwartz and Wendy B. Pratt

Holistic Resource Management Workbook
By Allan Savory

In Praise of Nature
Edited and with essays by Stephanie Mills

The Living Ocean: Understanding and Protecting Marine Biodiversity
By Boyce Thorne-Miller and John G. Catena

Natural Resources for the 21st Century
Edited by R. Neil Sampson and Dwight Hair

The New York Environmental Book
By Eric A. Goldstein and Mark A. Izeman

Overtapped Oasis: Reform or Revolution for Western Water
By Marc Reisner and Sarah Bates

Permaculture: A Practical Guide for a Sustainable Future
By Bill Mollison

Plastics: America's Packaging Dilemma
By Nancy Wolf and Ellen Feldman

The Poisoned Well: New Strategies for Groundwater Protection
Edited by Eric Jorgensen

Race to Save the Tropics: Ecology and Economics for a Sustainable Future
Edited by Robert Goodland

Recycling and Incineration: Evaluating the Choices
By Richard A. Denison and John Ruston

Reforming the Forest Service
By Randal O'Toole

The Rising Tide: Global Warming and World Sea Levels
By Lynne T. Edgerton

Saving the Tropical Forests
By Judith Gradwohl and Russell Greenberg

Trees, Why Do You Wait?
By Richard Critchfield

War on Waste: Can America Win Its Battle with Garbage?
By Louis Blumberg and Robert Gottlieb

Western Water Made Simple
From *High Country News*

Wetland Creation and Restoration: The Status of the Science
Edited by Mary E. Kentula and Jon A. Kusler

Wildlife and Habitats in Managed Landscapes
Edited by Jon E. Rodiek and Eric G. Bolen

For a complete catalog of Island Press publications, please write:
Island Press, Box 7, Covelo, CA 95428, or call: 1–800–828–1302